Rethinking Peace and Conflict Studies

Series Editor
Oliver P. Richmond
University of Manchester
Manchester, UK

This agenda-setting series of research monographs, now more than a decade old, provides an interdisciplinary forum aimed at advancing innovative new agendas for approaches to, and understandings of, peace and conflict studies and International Relations. Many of the critical volumes the series has so far hosted have contributed to new avenues of analysis directly or indirectly related to the search for positive, emancipatory, and hybrid forms of peace. New perspectives on peacemaking in practice and in theory, their implications for the international peace architecture, and different conflict-affected regions around the world, remain crucial. This series' contributions offers both theoretical and empirical insights into many of the world's most intractable conflicts and any subsequent attempts to build a new and more sustainable peace, responsive to the needs and norms of those who are its subjects.

More information about this series at
http://www.palgrave.com/gp/series/14500

Tarja Väyrynen

Corporeal Peacebuilding

Mundane Bodies and Temporal Transitions

palgrave
macmillan

Tarja Väyrynen
Tampere Peace Research Institute (TAPRI)
University of Tampere
Tampere, Finland

Rethinking Peace and Conflict Studies
ISBN 978-3-319-97258-9 ISBN 978-3-319-97259-6 (eBook)
https://doi.org/10.1007/978-3-319-97259-6

Library of Congress Control Number: 2018951039

This Palgrave Pivot imprint is published by the registered company Springer Nature Switzerland AG
The registered company address is: Gewerbestrasse 11, 6330 Cham, Switzerland

Acknowledgements

Various colleagues provided critical reactions to the book during its development. Thanks are owed to my research group on corporeal politics and movement (COMPORE). Eeva Puumala, Samu Pehkonen, Anitta Kynsilehto, Tiina Vaittinen and Elina Niinivaara created a stimulating and energetic environment that was central to the theorizing and conceptualization of the book. Our reading group and seminar with the Space and Political Agency Research Group (SPARG) was fundamental in thinking through issues related to space, agency and the political. I am also indebted to my colleagues and doctorate students at the Tampere Peace Research Institute (TAPRI) who have constructively commented the early versions of the chapters in our "article incubator" research seminar. I am extremely grateful to Eeva Puumala who allowed me to use our joint article as a basis for the third chapter and whose thinking on relationality has been inspirational.

I received assistance for the book from the Finnish Academy in the form of two large grants for The Body Politic of Migration and Mundane Practices of Peace projects.

This book is also about indeterminacy and transition, and thereby I am deeply indebted to wildlife and forests where I spend all my spare time with my dogs, Bertta and Kilu.

Contents

CHAPTER 1

Introduction: Corporeal Peacebuilding

Abstract The introductory chapter offers an elaboration of peacebuilding as corporeal event that becomes into being in mundane encounters. The objective is to argue that Peace and Conflict Studies can be revitalized by employing feminist theorizing, non-representational ethnography, post-colonial thinking and critical theorizing of everyday life. The chapter shows how taking the body seriously introduces phenomenological registers that prioritize the relational and vulnerable elements of human being and, thereby, mundane practices of peace-building and peace. In addition to that, the chapter urges a novel ethical stance to peacebuilding that is based on the Scandinavian research tradition where one of the initial goals was to study the ordinary mechanisms of conflict resolution and peace maintenance, not just violence and its management.

Keywords Body · Feminist theory · Everyday life

I seek to demonstrate in this book that peacebuilding, and consequently peace, is an event that becomes into being in mundane and corporeal encounters. In the process of developing my argument, I draw from several theoretical sources that include feminist Peace and Conflict Studies, non-representational ethnography, post-colonial thinking, critical theorizing of everyday life and feminist theorizing on the body and vulnerability. The book seeks also to establish a novel ethico-political

T. Väyrynen, *Corporeal Peacebuilding*, Rethinking Peace and Conflict Studies, https://doi.org/10.1007/978-3-319-97259-6_1

stance to peace, and in this sense return back to the early Scandinavian research tradition where one of the initial goals was to study the ordinary mechanisms of conflict resolution and peace maintenance, not just violence and its management (Jutila et al. 2008).

Furthermore, I seek to bring living and experiencing, sentient, body to the study of peacebuilding and peace. The book examines war and peace as socio-political institutions that begin and end with bodies. It, therefore, differs from the part of Peace and Conflict Studies where the human body is treated as an abstract and non-living entity. I demonstrate in the empirical chapters that war and peace touch our bodies in multiple ways and, as noted, the body is constitutive of war and peace. I illustrate through my empirical case studies also that different forms of violence render bodies vulnerable in different ways. I outline how bodies can resist violence, and argue that resistance is a precondition for peace. In other words, I show that vulnerable bodies that are not assumed to have agency indeed carry historical and political content as well disturbing and resistant qualities that are vital for the politics of peacebuilding, and ultimately for sustainable peace.

Through attending to witnessing, wounded, remembering, silenced and resistant bodies, the empirical cases of the book attest to the scope and diversity of conflict, peace and the political of post-conflict peacebuilding. My case study examples deal with home front, distant warfare, post-conflict affective landscapes and colonial violence. I employ material from Finland, but by discussing it from the perspective of political theory I seek to problematize peacebuilding in more general. The criteria for the selection of the cases in the book are their "closeness" as the book also wants to demonstrate that post-conflict is not "there and then", but also "here and now". The case study examples discuss "close" and mundane and invite Peace and Conflict Studies to focus on the everyday interactions through which peacebuilding takes place. The empirical material employed in the book is embodied data in the sense that it bears witness the affective and corporeal elements of the cases and events examined (cf. Walkerdine 2010). My primary material is audiovisual (e.g. documentaries, images), textual (e.g. narratives in media, interviews) and "artistic" (e.g. art work, performances) and it is collected from a variety of sources. It is hence mainly mediated material, i.e. I did not myself engage in its face-to-face/direct collection, but I show how this type of mediated material can bring valuable insights to Peace and Conflict Studies.

Peacebuilding refers in this book to John Paul Lederach's (1997, 20) broad definition where peacebuilding is understood to be:

> [...] a comprehensive concept that encompasses, generates, and sustains the full array of processes, approaches, and stages needed to transform conflict toward more sustainable, peaceful relationships. The term thus involves a wide range of activities that both precede and follow formal peace accords. Metaphorically, peace is seen not merely as a stage in time or a condition. It is a dynamic social construct.

My take on peacebuilding holds similarities with Lederach's definition as it emphasizes the relational and the political nature of peacebuilding. As Lederach points out, peacebuilding and peace are dynamic socio-political constructs. However, in this book my aim is to go beyond his definition and demonstrate that they are more: peacebuilding and peace are dynamic corporeal events and, thereby, they should be approached from multiple ontological positions.

Multiple Ontologies and the Political of Peacebuilding

As indicated, the aim of the book is to bring human body to Peace and Conflict Studies, and through this inclusion discuss the problematique of peacebuilding. The book attempts to shed light on the issues that are seldom examined in the field—e.g. trans-generational memory, resistance, silence, post-conflict governance of subaltern bodies, agency, the affective dimensions of war and conflict—from corporeal angles. When discussing the body, I argue that there is no singular body. Rather, body is always already multiple. This view derives from the above-mentioned conviction that there is no singular ontology; there are rather ontologies. The multiple ontologies of the body and, consequently, multiple bodies derive from the variety of forces that make bodies known to us. As Karen Barad (2003, 822) argues, there is a large number of material-discursive forces, including ones that get labeled "social", "cultural", "psychic", "economic", "natural", "physical", "biological", "geopolitical", and "geological", that are important to the processes through which we come to know the body. Barad's idea of forces enables me to see the "body multiple" (Mol 2002) through different ontologies and boundary-making practices that are formative of matter and meaning, productive of, and part of, the bodies that come into being. The book thereby attests to the variety of ontologies and forces and each chapter presents an unique perspective to the forces through which the human body becomes to be known. Ultimately, the aim to explore the implications these multiple ontologies have for peacebuilding and peace.

Since ontology is not singular and given, the question arises about which reality to "do" as there are always specific relevant interventions to be made. In this view, ontologies do not precede or escape politics, but have politics of their own. The question is about a politics of what is the reality that takes shape and that various people come to live with. This approach offers a novel perspective to peacebuilding as it deals with the corporeal reality of peacebuilding. Rather than asking whether or not some ontologies know the body correctly, the book shifts to the question what are the topics, the concerns and the questions that these ontologies bring forth in relation to the body, conflict resolution and peace, i.e. corporeal peacebuilding. Furthermore, I seek to show that different ontologies are related: since there is more reality than one, its different versions are variously entangled with one another. It is in this sense the book forms a collage (Sylvester 2009) that allows me to look a phenomenon from multiple angles, each of which offer a slightly different, yet complementary, understanding of peacebuilding.

This book is also about peacebuilding agency. It demonstrates that agency is about the abilities and the potential of the material, relational and vulnerable body to enact the "political" anew, not so much about certain subject positions. Agency is, in this vein of thought, about visibility and "*response-ability*" (Barad 2012, 206–207), about the possibilities of becoming visible and of mutual recognition, response and acknowledgement which is not to deny, but to pay attention to power and practices of governance. I outline in this book how the body's agentive capacity and political agency emerge in unexpected events, encounters and take multiple forms (cf. Väyrynen et al. 2017). Hence, the book contributes also to the literature that engages with the political potential of bodies. This requires me to move from the verbal to the embodied terrain of agency and action. I derive largely from non-representational theorizing on agency where the body's relationality is seen to be constitutive of its agency. Whilst in this tradition, the book seeks to cultivate an affinity for the analysis of "events, practices, assemblages, structures of feeling, and the backgrounds of everyday life against which relations unfold in their myriad potentials" (Vannini 2015, 317). In order to cultivate this type of affinity, the case study examples focus on events, practices, affects and the formations of everyday life that are corporeal, mundane and, thereby indeed relevant also to peacebuilding.

Bringing body and corporeal agency to Peace and Conflict Studies means bringing political theory to the field. It is important since political theory helps thinking through such issues as subordination, memory politics, temporality and spatiality. I am particularly interested in Jacques Rancière's theorizing as his theory of the political enables to discuss how the political is vital for peacebuilding. For Rancière, politics is not a matter of what people receive or demand nor is it a matter of the institutional creation of just social arrangements as a large part of peacebuilding theorizing argues. Rather, it is a matter of who is recognized to belong to the community and who is cast out. I consider this to be fundamental to the creation of sustainable peace, and therefore I examine the preconditions for the inclusion through the variety of case study examples. At the core of politics, according to Rancière (1998, 2009), there is *dissensus* or a break with the established order when the question "who can belong to the community" is asked. I argue in this book that Peace and Conflict Studies should not shy away from the study of dissensus as it is a precondition for peacebuilding and peace.

While engaging with dissensus, the book critically examines our ways of knowing peace. It studies the knowledge claims related to peacebuilding, peace and war, namely who has the right to speak about them, whose voice is acknowledged and heard. In this sense, the book is also about epistemologies of peace and about the creation of alternative ways of knowing. The focus on the epistemologies of peace implicates the existence of the multiple forms of violence that render people subordinate and silent. By discussing violence, I take part again in the Scandinavian Peace and Conflict Studies tradition where the discussion on violence has had a central place. The tradition has been dominated by Johan Galtung's distinction between overt and structural violence. His innovation was to argue that there is a form of violence that is embedded in social structures, where some social structure or social institution harms people by preventing them from meeting their basic needs (Galtung 1969, 1985). Although this innovation is still highly relevant when, for example, the multi-dimensionality and sustainability of peace is studied in post-conflict societies, the notion does not capture the complex societal mechanisms that render groups of people subordinate during peacebuilding nor does it reveal how subordination is transferred across temporal and spatial boundaries. Therefore, this book also broadens the notion of violence in order to capture the subtle ways (post-conflict) subordination and repression is produced and transferred across temporal and spatial boundaries.

Transitions, Choreographies and Events

The book is also about transitions, about the processes and periods of changing from one state or condition to another. I focus on transitions from war to peace, transitions as they take place in bodies, transitions across temporal and spatial dimensions, transitions in relations and transitions in the agency of the actors. The notion of transition enables me to emphasize peace as an ongoing process where multiple transitions are necessary.

In order to establish a link between the body and the transitive nature of everyday life and post-conflict agency, I have developed with my colleagues the notion of choreography in the context of mundane encounters. We argue that through the notion of choreography it is possible to capture the affective potential of everyday encounters as they articulate a body's capacity to communicate and integrate with other bodies. For us, hence, choreography is always social and connecting and it implies non-linear embodied relationality, extending and reaching out towards others, often across temporal and spatial boundaries. Because of its common association with dance, choreography is traditionally understood primarily to signify composing and arranging dances in advance, but for us, and in this book, it is something else: it is a force of affective potential as it articulates the above-mentioned body's capacity to communicate and integrate with other bodies (Väyrynen et al. 2017; see also Pehkonen 2016; Puumala and Pehkonen 2010; Puumala et al. 2011). We (Väyrynen et al. 2017, 11) summarise the idea of choreography:

> Through choreographies, we examine the oscillation of bodies as lived, experienced, and material configurations, which are simultaneously extremely personal, shared, and relational. The body is always partially marked and "already-made-visible", yet simultaneously always in the process of becoming, with other bodies in particular, historically contingent choreographies.

Choreographies do not happen in a vacuum, but they are enacted in the corporeal practices that people deploy in the everyday. Choreographies are open to surprises, and even disturbing, and they can produce extraordinary acts out of the ordinary (Väyrynen et al. 2017). In short, choreography in this book indicates practices of being (corporeally) in space and co-inhabiting space. Corporeal choreographies are situationally enacted in events as the everyday is made and remade thorough the changing positions and relations of bodies (cf. Vannini 2015, 320). Choreography therefore offers a non-conventional way of thinking about the emergent and constantly reconfigured entanglements of bodies, places, spaces and identities as well as of matter and discourse.

The research agenda promoted in this book, namely the study of corporeal peacebuilding, calls forth the analysis of choreographies and mundane micro-practices of peacebuilding. I concur with Robert Latham (2003, 1902) who argues that social scientists are "recognising the need to acknowledge the event-ness of world, along with the profound importance of affect in the unfolding of this event-ness". Latham's view comes close to Michael Bakhtin's idea of the event-ness of world where being itself is an ongoing event. Bakhtin's idea of Being as an event that has to be lived through holds, in my view, relevance for thinking about peacebuilding since it insinuates the research to focus on the rhythms and textures of everyday life, namely on life that comes into being though events and engages the human being in practical doing (cf. Boulding 1996). The event-ness of the world is relational as the self and other co-mingle in the event of being but remain distinctly "incarnated" (Gardiner 2000, 47). I draw from this tradition of thinking and focus on the event-ness of peacebuilding and peace, namely the capacity of co-being in everyday events as constitutive of peacebuilding.

I consider event to be a useful notion as events can be seen to have relatively vague spatial and temporal boundaries. They are in a state constant transition as they cross temporal and spatial boundaries. My take on event bears similarities with Alain Badiou's (2005) notion of event where events are gateways to future possibilities and reconfigurings by enabling novel modes of being-in-the-world. Badiou is sceptical towards understanding anything in the domain of the state as enabling because the state tends to keep its grip on the subject. Politics within the frame of the state is, according to Badiou, only a matter of reactive protests which do not produce reconfigurings of the social reality. Ultimately for Badiou, events are rare revolutionary moments in history. In this book, however, events are everyday occurrences that indeed open up novel ways being-in-the-world, not grand revolutionary moments. The case studies of the book can be read as examples of events that rupture normality and, at the same time, open a space to re-think peacebuilding arrangements and future possibilities for peace. Thinking through event-ness and the choreographies of events can gear, in my view, Peace and Conflict Studies towards the fleshy and carnal existence from which the mundane practices of peace emerge.

Through an attention to the specifics of particular events and an attention to detail, I seek to shed light in this book why mundane activities and encounters matter for peacebuilding. Latham (2003, 1903) discusses the focus on the events and the type of research that emerges:

> To take the event-ness of the world and affect seriously requires a shift from an empirical mode that is driven by the imperative to denote, to one oriented towards the work of description. The aim is no longer to seek after explanations that claim to go beyond the event being described, but simply to present descriptions that are infused with a certain fidelity to what they describe.

Although the focus in my case study examples is on a range of individual and collective embodiments, ultimately these accounts are about rather more than just singular body. As Latham (2003, 1903) notes, it is important that accounts of events "are also about how institutions function and operate, about how belief and ethos bring the world to movement, and about how the materiality of the world is profoundly bound up with these affective elements".

Diffractive Reading of the Material

In order to understand the work of power, institutions, bodies and affective dimensions of war and peace, I use a "diffractive reading" (Barad 2014) strategy where reading new insights are built by carefully reading for differences that matter in their details, together with the recognition that at the core of the analysis is ethics. According to Evelien Geerts and Iris van der Tuin (2016), reading and thinking diffractively "steps out of the phallogocentric, reflective logics of producing the Same all over again by acknowledging the differences that exist". While doing diffractive reading I seeks to point out where the problematic reductions and assimilations of difference have taken place. It is important to notice that this is neither a method nor methodology in the traditional sense of the word where research method implies a specific research design and procedure in order to answer a set of pre-given research questions. For me a method would do what Geerts and van der Tuin caution about, namely produce sameness that does not acknowledges differences. Rather than employing a hierarchical methodology that would put different texts, theories, and strands of thought against one another, diffractively engaging with texts, intellectual traditions and empirical material means that they are dialogically read "through one another" to engender unexpected outcomes (Barad 2007, 93). By using diffractive reading strategy, I become engaged with different disciplinary practices, and by blurring the boundaries between different disciplines and theories, my aim is to provoke new thoughts and theories and examine how they can be (re)made to matter more toward inclusion than exclusion and boundaries (cf. Geerts and van der Tuin 2016).

Why acknowledging differences is so important in the context of peacebuilding? I seek to show that peacebuilding and peace are founded on the creation of differences and that this has effects on post-conflict subjects and their bodies. In short, diffractive reading strategy reveals particular effects war, violence, peacebuilding and peace have on subjects and their corporeal being in the world. According to Donna Haraway (1997, 273) using this type of reading gives the opportunity to become more attuned to how differences are being created in the world and to examine the effects they have. Since a part of the ethos of Peace and Conflict Studies has always been active engagement with the world, diffractive reading is well suited as seeing and thinking diffractively implies also a self-accountable, critical, and responsible engagement with it. When peacebuilding is studied as an embodied and mundane phenomenon and when the focus of the study is on the body and body's choreographies, a critical and response-able reading is necessary.

When I use narratives as empirical material and employ diffractive reading strategy, it is important to note that narratives as such do not form the data of my study in a way that I would seek to produce knowledge claims on their truthfulness or accuracy. Rather, for me experiences and narratives/stories about experiences are socially constructed, and always in need of interpretation. The complexities of the usage of the term "experience" need also to be taken into account in this instance. In this book, experience refers both to *Erlebnis* (moments of sensation that are immediate and fragmented) and *Erfahrung* (cumulative and more sustained texture of experience). Experience is singular and transitory, but can produce a biographical closure when retrospectively reconstructed (Elsaesser 2009, 292–312). Catrina Brown (2013) concurs with the above-mentioned view when she writes that "while experiences are important as the juncture in which lived events and the corresponding meaning made of those events are joined, they are not inherently 'true' or 'valid'". Since they are not inherently true or valid, they require, in my view, deconstruction as well as diffractive reading. In this process the researcher becomes an active rather than passive participant in research when she provides the diffractive reading of the material. Narratives do not appear in void as all narratives are embedded in practices of power, social interaction, culture, and history, and thereby there are always multiple entangled voices and forces at work that constitute them. By engaging with the voices and forces, it is possible to recognize that narratives are always partial and situated (see Butler 1990a, b, 1992; Haraway 1988; Wibben 2011).

I employ also documentary films as my research material. They are often discussed in terms of knowledge claims, namely in terms of their truthfulness, accuracy and the "artistic freedoms" the director has taken. Contrary to this view, Rens van Munster and Casper Sylvest (2015) note that documentaries offer ways of seeing, of rendering things visible, and thereby they can challenge meaning we give to practices and representations. In this book I am not interested in the truth claims of films, but rather in the corporeal performances that make post-conflict differences visible. Although I am interested in knowledge claims and epistemologies of peace, for my purposes acknowledging the performative and corporeal power documentaries can have as they transfer affects across material, spatial and temporal boundaries is important. Since I write in the tradition of non-representational theory, documentary films provide me with images that have affective force. I derive inspiration from theorists such as Alan Latham and Derek McCormack (2009, 253) who argue that the force of images is not just representational, but images are also "blocks of sensation with an affective intensity: they make sense not just because we take time to figure out what they signify, but also because their pre-signifying affective materiality is *felt* in bodies". In short, I read documentary films (and art) in this book as events that have affective, corporeal and felt force.

My interest in arts and its relevance to Peace and Conflict Studies, on the other hand, relates to visibility, namely the capacity of art to make things visible. I rely particularly on Rancière's thinking and his understanding of social order as the "distribution of sensible". He (Rancière 2006, 12–13) writes:

> I call the distribution of the sensible the system of self-evident facts of sense perception that simultaneously discloses the existence of something in common and the delimitations that define the respective parts and positions within it. A distribution of the sensible therefore establishes at one and the same time something common that is shared and exclusive parts. This apportionment of parts and positions is based on a distribution of spaces, times, and forms of activity that determines the very manner in which something common lends itself to participation and in what way various individuals have a part in this distribution.

According to Rancière (2010), art is a challenge to a particular partition of the sensible as sensibility itself becomes a field of artistic creation and experimentation. Art is field of experience that is severed from its traditional reference points, and as such open for new restructurings. Art can hence reveal the contingency of a particular partition of the sensible by constructing another one. In this view art and politics are closely related: both of them can reveal the contingency of the existing social order. Rancière (2010, 140) writes that "art and politics each define a form of dissensus, a dissensual re-configuration of the common experience of the sensible". Furthermore, art re-frames the world of common experience as the world of a shared impersonal experience. In this way, it aids to help create the fabric of a common experience in which new modes of constructing common objects and new possibilities of agentive capacities may be developed (Rancière 2010, 140). Rancière, hence invites us to examine the distribution of the sensible in the social field, investigating what is visible and invisible in terms of, for example, public discourse and various social identities. These are all ultimately questions of new social and political formations, identifications and agency. The question in this books then becomes one of how new forms of post-conflict being-in-the world that depart from the old order might be produced through peacebuilding. In the domain of epistemology and ontology, the question is no longer that of the nature of reality, but rather of the distribution of the sensible, and thereby belonging, within which we find ourselves immersed in post-conflict situations.

Organizing the Book

My argument is developed in the following seven of chapters. In chapter (Chapter 2) I further develop the conceptual frame of the book. My aim is to introduce a phenomenological register that involves a move from totalizing perspectives to micro-practices and mundane practices where lived experience offers a rich fabric of corporeal presence, relationality and affect. I argue in the chapter that the radical and transformative aspects of everyday life can be examined by exposing the extraordinary in the seemingly ordinary, and therewith the radical potential embedded in the everyday for peacebuilding revealed. I establish the foundations of the book on feminist Peace and Conflict Studies where the relationality of human existence is emphasized. In other words, to elaborate the importance of the body and its relationality for the study of peacebuilding, the book turns to feminist theorizing which has a strong commitment both to the corporeality and to the everyday. I argue that the mainstream theorizing of peace has dismissed this tradition and remained in the abstracted notions of peacebuilding and peace. By drawing from the early feminist Peace and Conflict Studies, feminist theorizing on vulnerability and critical theorizing on everyday, Chapter 2 paves the way to empirical chapters which examine sensuous, embodied, non-cognitive, pre-intentional and commonsensical of the everyday life and lived experience of conflict, violence and peacebuilding.

The third chapter (Chapter 3) of the book engages with the lived body and personal histories of war experience. It uses the stories of two old ordinary Finnish women, Ada and Bertha, who experienced the Second World War. I seek to illustrate how people even in harsh conditions and unexpected situations find innovative, and very mundane, ways of being in the world and relating to other bodies. The chapter discusses sensuous experience of war and peace that defies the simplistic categorizations and interpretations of them. My aim is to introduce an alternative onto-epistemological take on Peace and Conflict Studies so as to overcome the separation of the discipline from ordinary people and mundane practices. The chapter relies on a notion of the relational body put forward by early feminist peace researchers and Jean-Luc Nancy, where the body refers to the relational foundation of the "being-in-the-world" . The chapter lays the groundwork for the rest of the book by making visible the lived experience of ordinary persons. By focusing on bodies, on the other hand, and particularly on bodies coming together the chapter enables also the study of the war and peace in ways that challenge the prevailing and normalized assumptions about conflict and peacebuilding agency.

In Chapter 4, I cultivate new materialist ontology and examine again how bodies are touched by war and violence. Although for most of the western population battlefields are remote, we indeed encounter war and violence in the form of war veterans' bodies. In other words, veterans' physically and mentally mutilated and disfigured bodies bring the war and violence to us. Through the study of Finnish Afghan war veterans' presence in the Finnish affective landscape, the chapter shows how the distant warfare comes close to us and how the non-representational transference of war takes place. In addition to the lived experiences of the Finnish Afghan war veterans, the focus in this chapter is on how the touch of violence is transferred across time and space and how this touch makes peacebuilding a long-term affective and corporeal process. In this chapter my aim is to bring a new angle to the body and peacebuilding through the work of Karen Barad (2007, 2012) and her new materialist ontology. Her views are complemented with Erin Manning's (2007) theorizing which provides a means to understand affects that emanate from the materiality of human bodies. In sum, in this chapter I examine how distant warfare touches us, how warfare is brought home to us through touch and what kind of ethical implications the touch of violence has for peacebuilding.

In Chapter 5, Elma, Roosa and Terttu narrate their experiences of the Second World War, and particularly of their escape from Finland with the Nazi army. In this chapter I focus on disciplined and silenced post-conflict bodies as well the potential for resistance embedded in these bodies. The chapter shifts the attention to the practices of governance that shape bodies during war and post-conflict peacebuilding. I examine how bodies are controlled both during and after conflict by a variety of disciplinary practices and how bodies, however, can resist the practices of governance. This chapter also demonstrates how abjected bodies can carry historical and political content that is not reconciled with the post-conflict collective identity narrative. The chapter shows how the Finnish post-conflict national subject came into being through the establishment of differences and the other. I argue that the narratives of these women are not simple representations of a past event, but political performances and utterances which intervene in a seemingly unitary post-conflict peace and introduce dissensus to the core of the nation.

I introduce more vulnerable, yet resistant, bodies that do not yield to the post-conflict collective memory work and unitary post-conflict identity politics in Chapter 6. The chapter employs artist Seppo Salminen's, a son of a Second World War veteran, and Kaisu Lehtimäki's, a woman who was accused to have fraternized with German soldiers during the Second World War, public performances to show how they re-enact in novel ways the differences the war and violence had created. The chapter shows that the ways in which the past conflict is corporeally remembered are fundamental to peacebuilding as the practices of remembering war and violence bring forth tense locations and differences, namely the political in peacebuilding. In sum, in this chapter I expand the discussion on resistant bodies and examine how they can disturb post-conflict consensus by taking part in the distribution of the sensible and bring forth the political of long-term peacebuilding. I seek to outline how a remembering body can resist the disciplinary control as well as challenge peacebuilding which relies on linear national time, controlled collective memory-work and that seeks closures through harmonized and authenticated facts about the past violence in order to create a homogeneous post-conflict collective identity.

In Chapter 7, I draw from indigenous activism and art to demonstrate how they make visible the violent boundaries that distribute "our place and time" and "their place and time", namely the distribution of the sensible. The chapter discusses peacebuilding in relation to colonial violence, namely Finnish-indigenous Sámi settler-colonialism. I show how the political that emerges from indigenous resistance guarantees that the colonial dominance erupts to the social life after hundreds of years of invisibility, and thereby, creates a precondition for peacebuilding and peace. The chapter demonstrates also how violence embedded in settler-colonial relations goes often unrecognized since it is slow, structural and epistemic in the sense that it functions through corporeal practices that seek to neutralize the differences, produces standardized patterns of being a citizen and render alternative forms of existence as well as histories and futures unrecognized. Identifying the variety of long-term practices that harm and do violence is often impossible exercise in the case of long-term epistemic and slow violence. I argue that transforming colonial violence is not merely a stage in time or a condition, but a dynamic long-term social construct as colonialism is a corporeal entanglement of political social and economic practices, ideologies, relationships, collections and archives.

In the concluding chapter (Chapter 8) I bring together the previous chapters and their ontologies without, however, offering a singular ontology. Furthermore, I expand the discussion to the premises of the political in peacebuilding and peace. The chapter argues on the bases of the empirical case study examples that peacebuilding, post-conflict pluralistic society and peace is not about the creation of a consensus, but about the distributing of the sensible, of becoming visible, and thereby about dissensus. I engage there again with political theory and the chapter is written in a form of research agenda that promotes response-able peacebuilding.

References

Badiou, A. (2005). *Being and Event*. London and New York: Continuum.

Barad, K. (2003). Matter Feels, Converses, Suffers, Desires, Yearns and Remembers. *Signs: Journal of Women in Culture and Society, 28*(3), 801–831.

Barad, K. (2007). *Meeting the Universe Halfway: Quantum Physics and the Entanglement of Matter and Meaning*. Durham: Duke University Press.

Barad, K. (2012). On Touching—The Inhuman That Therefore I Am. *Differences, 23*(3), 206–223.

Barad, K. (2014). Diffracting Diffraction: Cutting Together-Apart. *Parallax, 20*(3), 168–187.

Boulding, E. (1996). Peace Behaviours in Various Societies. In *From a Culture of Violence to a Culture of Peace* (pp. 31–53). Paris: Unesco.

Brown, C. (2013). Women's Narratives of Trauma: (Re)Storying Uncertainty, Minimization and Self-Blame. *Narrative Works 3*(1). Accessed 28 May 2018. https://journals.lib.unb.ca/index.php/NW/article/view/21063/24305.

Butler, J. (1990a). *Gender Trouble: Feminism and Subversion of Identity*. New York: Routledge.

Butler, J. (1990b). Gender Trouble, Feminist Theory, and Psychoanalytic Discourse. In L. Nicholson (Ed.), *Feminism/Postmodernism* (pp. 324–340). New York: Routledge.

Butler, J. (1992). Contingent Foundations: Feminism and the Question of Postmodernism. In J. Butler & J. Scott (Eds.), *Feminists Theorize the Political* (pp. 3–21). New York: Routledge.

Elsaesser, T. (2009). Between Erlebnis and Erfahrung: Cinema Experience with Benjamin. *Paragraph, 32*(3), 292–312.

Galtung, J. (1969). Violence, Peace and Peace Research. *Journal of Peace Research, 6*(3), 167–191.

Galtung, J. (1985). Twenty-Five Years of Peace Research: Ten Challenges and Some Responses. *Journal of Peace Research, 22*(2), 141–158.

Gardiner, M. (2000). *Critiques of Everyday Life*. New York and London: Routledge.

Geerts, E., & van der Tuin, I. (2016). Diffraction & Reading Diffractively. *New Materialism*. Accessed 28 May 2018. http://newmaterialism.eu/almanac/d/diffraction.

Haraway, D. (1988). Situated Knowledges: The Science Question in Feminism and the Privilege of Partial Perspective. *Feminist Studies, 14*(3), 575–599.

Haraway, D. (1997). *Modest_Witness@Second_Millenium. FemaleMan©_Meets_OncoMouse™: Feminism and Technoscience*. New York: Routledge.

Jutila, M., Pehkonen, S., & Väyrynen, T. (2008). Resuscitating a Discipline: An Agenda for Critical Peace Research. *Millennium: Journal of International Studies, 36*(3), 623–640.

Latham, A. (2003). Guest Editorial. *Environment and Planning, 35,* 1901–1906.
Latham, A., & McCormack, D. (2009). Thinking with Images in Non-representational Cities: Vignettes from Berlin. *Area, 41,* 252–262.
Lederach, J. P. (1997). *Building Peace: Sustainable Reconciliation in Divided Societies.* Washington, DC: U.S. Institute of Peace Press.
Manning, E. (2007). *Politics of Touch.* Minneapolis: University of Minnesota Press.
Mol, A. (2002). *The Body Multiple: Ontology in Medical Practice.* Durham: Duke University Press.
Nancy, J.-L. (2000). *Being Singular Plural.* Stanford: Stanford University Press.
Pehkonen, S. (2016). Choreographing the Performer-Audience Interaction. *Journal of Contemporary Ethnography, 43*(3), 306–330.
Puumala, E., & Pehkonen, S. (2010). Corporeal Choreographies: Failed Asylum-Seekers Moving from Body Politics to Bodyspaces. *International Political Sociology, 4*(1), 50–65.
Puumala, E., Väyrynen, T., Kynsilehto, A., & Pehkonen, S. (2011). Events of the Body Politic, A Nancian Reading of Asylum-Seekers' Bodily Choreographies and Resistance. *Body & Society, 17*(4), 83–104.
Rancière, J. (1998). *Disagreement.* Minneapolis: University of Minnesota Press.
Rancière, J. (2006). *The Politics of Aesthetics.* London: Continuum.
Rancière, J. (2009). A Few Remarks on the Method of Jacques Rancière. *Parallax, 15*(3), 114–123.
Rancière, J. (2010). *Dissensus: On Politics and Aesthetics.* London: Continuum.
Sylvester, C. (2009). *Art/Museums: International Relations Where We Least Expect It.* Boulder: Paradigm Publishers.
van Munster, R., & Sylvest, C. (2015). Documenting International Relations: Documentary Film and the Creative Arrangement of Perceptibility. *International Studies Perspective, 16*(3), 229–245.
Vannini, P. (2015). Non-representational Ethnography: New Ways of Animating Lifeworlds. *Cultural Geographies, 22*(2), 317–327.
Väyrynen, T., Puumala, E., Pehkonen, S., Kynsilehto, A., & Vaittinen, T. (2017). *Choreographies of Resistance: Mobile Bodies and Relational Politics.* London and New York: Rowman and Littlefield.
Walkerdine, V. (2010). Communal Beingness and Affect: An Exploration of Trauma in an Ex-industrial Community. *Body & Society, 16*(1), 91–116.
Wibben, A. (2011). *Feminist Security Studies: A Narrative Approach.* New York and London: Routledge.

CHAPTER 2

Thinking and Theorizing Peace: Corporeal and Mundane Dimensions

Abstract The chapter shows how the early feminist peace researchers insisted on the importance of the relational body and mundane when studying conflict resolution, peacebuilding, reconciliation and peace, and how their research agenda has not been actualized in the mainstream theorizing of peacebuilding. Although feminist peace research has received a little acknowledgement among mainstream peace scholars this chapter shows that this radical scholarship is instrumental when investigating peacebuilding from corporeal perspectives. The chapter introduces also the "local turn" scholarship and argues that it is limited when it comes to its understanding of the "here and now" everyday and mundane practices of peace. An alternative understanding is offered, emerging from Judith Butler's observation that adequate responses to different forms of violence require the recognition of human vulnerability.

Keywords Mundane encounters and vulnerability · Feminist peace · Mundane practices of peacebuilding · Local turn

In addition to the body, this book attests, as indicted in the introductory chapter (Chapter 1), to the everyday as these two are co-constitutive in many ways. The book examines what the everyday is and what kind of ethico-political implications is has for peacebuilding and peace. I seek to demonstrate that peacebuilding ultimately actualizes through the everyday. Although the body is approached in the empirical case studies

T. Väyrynen, *Corporeal Peacebuilding*, Rethinking Peace and Conflict Studies, https://doi.org/10.1007/978-3-319-97259-6_2

through multiple ontologies, phenomenological register connects these ontologies as it directs the research to lived experience, to experience that is always already embodied and mundane. In line with phenomenological sociology the book gives a specific meaning to the everyday and thereby provides a starting-point for the critical understanding of the everyday. It relies on a branch of thought where everyday is the world of natural attitude and where many things are taken for granted. Everyday is the intersubjective, corporeal and relational world where typifications through which we interpret the world and act in it are handed down to us (Schutz 1967; Väyrynen 2001). Phenomenological theory gears the focus of the research towards the intersubjective ways in which individuals manage social and cultural material within the context of everyday life (Väyrynen 2001, 2005).

Using the phenomenological register requires me to move from totalizing perspectives to micro-sociological approaches and, hence, mundane practices. It follows that the radical and transformative aspects of everyday life are examined in this book by exposing the extraordinary in the ordinary as also indicated in Chapter 1. Although the everyday life is the world of shared typifications and cultural material, it is also a world where the subject has "living presence" through and in her body. In the phenomenological tradition, for Edmund Husserl (1992, 20) the body is the zero point of reference of all our orientation, a point around which our world is centered. According to Maurice Merleau-Ponty (1962, 88–92), on the other hand, we gain access to the world through the body and, hence, our experience of the everyday depends upon a "lived body". In this vein of thought, the symbolic order is constituted through the body as the body's being-in-the-world is at once mediated through physical embodied presence and cultural meanings (Crossley 1995, 44–45). Our living presence as sentient beings and the apprehension of the symbolic actualizes in our bodies, or rather in our relational bodies as the body is always already in relation to other bodies.

Feminist Peace and Conflict Studies

In order to elaborate the importance of body and the everyday for the study of peacebuilding, I turn to feminist theorizing which has a strong commitment to both of them (Enloe 2006; Jenkins and Reardon 2007; Brock-Utne 1989; Tickner 1992; Boulding 1990, 1996; Ruddick 1990, 1993, 1995 , 2001). Furthermore, this corpus of thinking establishes a solid connection between corporeal ontology and peace epistemology, and as such provides a foundation for my case studies. Although feminist peace research has received a little acknowledgement among mainstream peace scholars (Vaittinen 2017), the corpus of this radical scholarship is instrumental when investigating the body, the ordinary and the everyday and their connection to knowledge-formation in post-conflict settings (for exceptions which fully recognize the feminist agenda see Confortini 2006; Confortini and Ruane 2013; Vaittinen 2017). There are multiple ontologies in the feminist peace theory, yet the concern for marginalization and the understanding of the corporeal and relational nature of human existence is the key contribution that indeed enables a new take on the corporeality of peacebuilding as well as on the spatial and temporal transitions that are a part of peacebuilding.

In the feminist peace theorizing, Elise Boulding' (1996; cf. Väyrynen and Puumala 2015) focus on peace as a daily process returns peace back to the very ordinary and its corporeal rhythms. Her work emphasizes the personal and interpersonal promotion of peace as for her it involves shaping and reshaping understandings and behaviors to adapt to a constantly changing world and sustain well-being for all. Boulding's theory is instructive for my book as it is about praxis of peace where the daily doing peace is fundamental. In addition to its emphasis on the daily doing of peace, Boulding's views concur with the critical thinking on the everyday where it is argued that our existence in the world and our everyday life is embodied and relational, i.e. human existence is based on our dependency on the others and it is hence always multiple. Her view resonates with Maurice Merleau-Ponty's (1962, 19, 173) notions of "intercorporality" and "carnal intersubjectivity" which denote a primordial carnal bond between human beings. According to this view, we are always open to each other, and with others. Boulding's theorizing, in my view, provides a new opening where radical peacebuilding theory does not limit itself, for example, to a civil society and its resistant and emancipatory potential. Rather, radical theory, such as Boulding's, is "attuned to all facets of human existence: the poetic, irrational, corporeal, ethical and affective" (Gardiner 2000, 19).

Sara Ruddick (1995, 219) elaborates further in the Bouldingian spirit the relational nature of peace. She argues that peace is a matter of creating relationships in which people experience their being respected. At the core of Ruddick's theorizing is her thinking about motherly care that is the opposite of the practices of violence and war. Care is always a relation, and as such it is also a foundation for knowing the world: we become to know the world through corporeal, mundane and relational practices, by being an embodied part of the world. According to Ruddick (1990, 1995), the relational practices of mothering and the knowledge generated through them advance skills in peacemaking (cf. Robinson 2011; Vaittinen 2017).

Ruddick's way of connecting care and our ways of knowing establishes a link between corporeal ontology and epistemology. Catia Confortini and Abigail Ruane (2013, 97) expand her theorizing by arguing that mothering and the knowledge generated through it engage us in the practices of de-centering ourselves. De-centering implies both locating oneself into a situation and, at the same time, stepping out from it. This double move enables us to hold someone else in personhood. Confortini and Ruane (2013, 71) write that we need "participatory epistemology, or an understanding of how actors develop knowledge through their practices of engagement with others". Knowing in this view is not a skill of a detached, unembodied, observer, but grows from the interactions between multiple beings and bodies, living together and acting together in a world that is in a process of constant becoming (Confortini and Ruane 2013, 80).

In addition to care being a "motherly practice" care is seen to be in this book an "intercarnal relation", a practice of engagement with a corporeal other. I concur with Michael Gardiner who argues that care is ultimately an ordinary gesture of the everyday, the unspoken desire of the body and a microscopic expression of solidarity, "where the predemptive promise of the everyday continues to persist, in the interstices of more formal social relations and organizational structures" (Gardiner 2000, 17). However, everyday is not solely the realm of solidarity, but also the distribution of violence, torture and massacres haunt and shape the everyday and its relations (Das 2007). By drawing from the above mentioned literature and by paying attention to sensuous, embodied, non-cognitive, pre-intentional and commonsensical of the everyday life and lived experience, the empirical chapters of this book examine peacebuilding in relation to haunting violence, acts of micro-solidarity and the epistemological potential of the everyday. In other words, attention is paid to events, affective states, what is left unsaid as well as the incompletedness of the corporeal everyday performances (cf. Dewsbury 2010; Vannini 2015, 319). As such, the book forms a critique of thinking peacebuilding and peace in abstract, theoretical and philosophical terms.

Mainstream Theorizing: From Abstracted to the Local Turn

How feminist Peace and Conflict Studies is located in relation to the mainstream thinking of peace and how it differs from it? Unlike in feminist Peace and Conflict Studies, peace in Philosophy, International Relations and mainstream Peace and Conflict Studies has been seldom contextualized, namely socially and politically situated. Rather, it has been abstracted from its everydayness, and therefore from its visibility too. Peace has become an elusive concept which has been deployed to "bludgeon humanity with its extraordinariness, forever out of reach, illusive by definition, a dream too flatteringly sweet to be substantial" as Rose Mary Shinko (2008, 489) describes the problematique of "abstract peace". In empirical research peace is often thought to be the opposite of the causes of war and peacebuilding is seen to be a removal of these causes. However, as Séverine Autesserre (2014, 495) notes, it is important to ask "why certain places experience little violence", and even more importantly "what are the sources of peace" as this could open up new ways of thinking war and peace. The Everyday Peace Indicators project (Firchow and Mac Ginty 2017) that seeks to map the bottom-up indicators of peace is a step towards the direction of examining the sources of peace, and yet its focus on "subaltern peace" is too narrow for my agenda as I will argued later in this chapter. Drawing from this vein of thought that focuses on peace, I seek to argue in this book that peacebuilding can be studied in ways that does not remove it "out of reach" nor dismiss the sources of peace. In doing this I engage actively with the heterogeneous entanglements of the corporeal and mundane practices of peacebuilding and peace as the feminist thinking on peace suggests.

Although the mainstream theorizing on peace has developed in close interplay with social interests, the interests do not constitute an integral part of the research. Particularly the post-Second World War *Sorge* (Heidegger 1996, 175–180), "concerned mood", for the future of the globe and the realization that mass destruction is possible, has guided the mainstream theorizing on peace. In the field of International Relations' (IR) such realist scholars as Raymond Aron, Hans Morgenthau and George Kennan engaged in the theorizing of peace within the power political frame and status quo whereas so called English school saw there are rules that are constitutive of the global society, and thereby restrain the use of force (Richmond 2001, 2010). The first peace researchers challenged the existing International Relation's thinking on peace by arguing that the power political and state-centric view of the world remains distorted as it does not notice the different forms of violence, the multiplicity of actors in the world politics and the cob-web type of connections between the actors (Wallensteen 2011).

In addition to the widening of the agenda—and whilst dismissing the existing praxis oriented feminist agenda—many first generation peace researchers argued that so called scientific thought form and, for example, abstract modeling would guarantee a scientific theory of peace. Unlike the Anglo-American Peace and Conflict Studies tradition, the ethos of Scandinavian research tradition soon developed to the direction of empiricism which formed a reaction against dogma of presentism and factuality as well as against overtly abstract scientific frames found in, for example, in the realist school of IR. The Scandinavian peace thought argued that single-minded focus on the present "state of affairs" should be complemented with an open time horizon where peace is theorized by "integrating empirical studies of the past, critical studies of the present and constructive studies for the future" (Galtung 1985, 141). This ideal functioned as a critique towards the American Peace and Conflict Studies as it was represented, for example, in Journal of Conflict Resolution and which was dominated by quantitative research designs (Jutila et al. 2008).

Oliver Richmond's (2010) categorization of the generations of peacebuilding is instructive for summarizing the mainstream peacebuilding and peace theorizing. Although Richmond's categorization derives from the practices of peacebuilding, it captures also the key concepts of theorizing. The first generation theorizing focused on the negative peace and concentrated on power sharing and balance of power arrangements within and between the states. As noted above, peace was in this vein of thought theorized in terms of the possibility of a status quo and balance of power. In the second generation, peace was conceptualized in terms of institutional and legal designs where collaboration between a range of international actors were seen to guarantee peace. The third generation approach circulated around the notion of a peace built with international norms, law, and institutions, and the liberal democratic and "marketized" concept of the state. The third generation coincides with the democratic peace theory where democracies are argued to be hesitant to go to war against each other.

This book clearly departs company from the above-mentioned generations and agendas in Peace and Conflict Studies as it suggests a research agenda which resonates with the early feminist peace thinking introduced above and whose ambition, thereby, is to examine the corporeality of peacebuilding and to re-theorize peacebuilding and peace in ways which defy the non-situatedness and abstractness of it. The book locates peace in social and political contexts, and examines the practices and event-ness of peacebuilding. My aim in this book is to demonstrate that the micro-sociology of corporeality and relatedness enable a renewed grasp of the study of peacebuilding. My research agenda hence radically differs from the mainstream abstract or "peace-as-opposite-of-violence" theorizing on peace by locating the substantialness and situatedness of peace within political and social life. The book is hence a substantive study of particular aspects of peacebuilding, but my aim is also to be theoretically ambitious: I seek to (re)think a counter-tradition in theorizing peacebuilding, one that has been largely ignored in the mainstream study of peace.

So-called "local turn in peacebuilding" scholars (Autesserre 2014; Björkdahl and Höglund 2013; Gearoid et al. 2013; Hohe 2002; Leonardsson 2015; Millar 2014; Millar et al. 2013; Mac Ginty 2008, 2011, 136–163; Mac Ginty and Richmond 2013; Randazzo 2016; Richmond 2009a, b, 2012; Richmond and Mitchell 2011) agree with the critique of abstractness and the lack of context of theorizing peace. In doing this they aim at dismantling the Western bias of both theorizing and practicing peacebuilding. The *Sorge* of this scholarship no longer relate to the possibility of mass destruction, but to the dominance of neo-liberal governance and its implications for peacebuilding which, according to many scholars writing in this tradition, seeks to reproduce and impose Western models and the reconstruction of "Westphalian" frameworks of state sovereignty. The liberal framework of individual rights, winner-takes-all elections and neoliberal free market economic programmes are seen to be at the core the models (Chandler 2010). In order to reveal the problems of liberal peace, Mac Ginty and Richmond (2013, 764) argue that peace is not single, namely neo-liberal as the neo-liberal institutions would prefer it to be. Rather, it is always hybrid where the local meets the international, and often contested since the form of peace implemented through the practices of neo-liberal governance is moulded in "local" contexts in resistant manners. In short, the imposition of the Western peacebuilding model is doomed to fail. It follows, according to Richmond and Mac Ginty, that peacebuilding is a practice where subjects exercise their "agency for peace or against it, doing their best to maintain a viable everyday existence in the face of governmentalism (global government, for example, in the name of liberal peace) and structural power (where power is exercised regardless of its implications for order or peace)" (Mac Ginty and Richmond 2013, 764).

For the "local turn" scholarship subaltern views of peace become important since subaltern actors possess everyday agency in relation to peace either by promoting it or resisting the top-down neoliberal peacebuilding attempts. The everyday is fundamental for this branch of thinking as the "pursuit of everyday tasks may allow individuals and communities in villages, valleys and city neighborhoods to develop common bonds with members of other ethnic or religious groups, to demystify 'the other' and to reconstruct contextual legitimacy" (Mac Ginty and Richmond 2013, 769). Given the agency the subaltern exercises, the research needs to focus on the everyday life, the forms of transversal solidarity and crass roots mobilization that enable peacebuilding and state-building to operate in successful ways. In other words, it is thought that vibrant civil society and its network type of everyday solidarity that cuts across ethnic and religious affiliations is the guarantee of peace. However, it is argued in the local turn that the promoters of liberal peace have difficulties in seeing this type of agency and solidarity as they focus on elite-lead and top-down approaches to peacebuilding, and thereby miss the opportunity for sustainable and local peace.

Although the "local turn" functions as an important critique of the neoliberal governance and peacebuilding, for my research agenda and for the purposes of this book where the aim is to capture the corporeality of peacebuilding, the discussion and understanding the everyday in the local turn literature is limited: a diffractive reading of it is needed. Everyday implies for me more than the "local" and "ordinary people" and their resilience and resistant potential targeted at neo-liberal peacebuilding as it largely understood to be in the local turn scholarship. From the perspective of my research the interest in the everyday in the local turn is too narrow and has not been translated into sustained considerations of the productiveness of mundane practices.

Where I most radically depart company from the "local turn" and its understanding of the everyday is its tendency to define everyday through the subaltern and her spatiality, namely through the spatial location where the subaltern is assumed to reside and through the forms of everyday life that are though to characterize her. In the local turn subaltern and her everyday life is projected to be somewhere "out there", in the distant location of the violent political conflict or post-conflict zone. This is, in my view, a traditional Orientalist projection where the spatially projected "local" world and its everyday are constructed to be radically different from our own world (cf. Paffenholz 2015, 861; Pieterse 2010, 62–82; Randazzo 2016, 1357–1361). In the projection the local and subaltern are constructed as distant others whose emancipatory potential the (Western and male local turn) scholar is concerned. This is a reminder of the colonial condition where white men are seeking to save "brown women from brown men" (Spivak 1994, 93). Although for example Roger Mac Ginty (2015, 841) suggests that there is need to move away from the overtly territorial understanding of peace, he does not develop the idea further. He writes that "if we understand the local as de-territorialised, networked and constituted by people and activity rather than place, then standard meanings of peace require reappraisal". What is implicated in the quote is that the local equals networked activity, transversal solidarity, rather than spatial location. As it will be argued later in the book, this does not yet capture the everydayness and eventness of peace.

In order to fulfill the gap in the literature, I turn in the case studies into the everyday and ordinary that are constituted through enactments, i.e. that are corporeal rather than spatial—and not necessary the "subaltern" in the Orientalist sense implying the non-Western populations. The focus of the book on embodied choreographies enables me to retain the local from the spatial hegemony of the distant other. Furthermore, this move implies, as noted earlier, the return to the one of the core ideas of Scandinavian Peace and Conflict Studies where the study of dispute management processes in peaceful contexts was seen to be a useful addition to the studies dealing with violent political conflicts (Jutila et al. 2008; cf. Autesserre 2014). Although this quest did not receive full acknowledgement at the time of its writing in the 1970s, it is, in my view, worth developing further.

Complementing the Early Feminist Agenda

As demonstrated, the early feminist peace researchers insisted on the importance of the relational body when studying conflict resolution, peacebuilding, reconciliation and peace, and that their research agenda has not been actualized as the mainstream theorizing has abstracted peace. As also argued, the "local turn" scholarship is limited when it comes to its understanding of the everyday and mundane practices of peace. In my view a diffractive reading of these corpuses of literature would produce a renewed research agenda. I short, they require more nuanced reading of power and governance as well as vulnerability they produce. My aim is to demonstrate through the case studies that body is always already both "active and acted upon" (Crossley 1995, 51) since the practices of governance that are targeted at the body produce it. Yet, the fleshy living body has agency that makes it "engaged body-subject" (Crossley 1995, 101). In other words, body is never a passive target of the practices of governance: it has a capacity to escape the practices (Väyrynen et al. 2017).

From the perspective of mundane peacebuilding developed in this book, Judith Butler's (2009) observation that one way of managing populations, i.e. biopower, is to distribute vulnerability unequally among people is crucial for my argument. Since biopower forms a norm which allows for measurement, evaluation and hierarchical ranking, and as such constitutes a mechanism of control, it distributes invulnerability and vulnerability among people. Power is in this sense relational as it establishes socio-historical relationships that render some bodies more vulnerable than others. Biopower increases efficiency and capacity at the level individual bodies and whole populations, and in order to do this practices of governance based on measurement, evaluation and hierarchical ranking distribute vulnerability and invulnerability. This is particularly acute during the times of economic, social and political crises, such as peacebuilding, when the social relations need to be re-imagined and re-configured.

In addition to the "governance induced vulnerability" that emerges from one being embedded in specific structures of power, there is also another type of vulnerability, namely vulnerability that is a condition of life (Butler 2004, 19; 2009, 30). As indicated earlier, body is constituted by discourse, including practices of power, but is not fully exhausted by it. In certain, for example, institutional contexts or extreme conditions the body's material qualities bring its vulnerability forth. As Louise Bethlehem (2013, 11) writes on the materiality and "deformation" that reveals the body's materiality: "the deformation of body" whether through for example violence "may momentarily release access a disruptive corporeality (at the border of bodily existence, as the borders of bodily existence) that gives up the ghost: the body's debtedness to materiality that cannot be thought in itself". Bethlehem's view can be interpreted to mean that at the border of bodily existence, e.g. during the processes of illness and dying, severe injury, disability, the material dimensions of existence unveil body's vulnerability to material forces.

Butler (2003) theorizes peace by arguing that peace is a commitment to living with a certain kind of vulnerability to others as vulnerability is a way of indicating one's dependency on another and to be acknowledged. She argues (2003):

> I think that peace is the active and difficult resistance to the temptation of war; it is the prerogative and the obligation of the injured. Peace is something that has to be vigilantly maintained; it is a vigilance, and it involves temptation, and it does not mean we as human beings are not aggressive. It does not mean that we do not have murderous impulses. This is a mistaken way of understanding non-violence. Many people think, "Oh, we need to be non-violent; humans can somehow get violence out of their souls; we're not constituted by aggression." Rather, I think it is precisely because we're constituted with aggression, it's precisely because we are capable of waging war, and of striking back, and of doing massive injury, that peace becomes a necessity. Peace is a certain resistance to the terrible satisfactions of war. It's a commitment to living with a certain kind of vulnerability to others and susceptibility to being wounded that actually gives our individual lives meaning. And I think this way of viewing things is a much harder place to go, so to speak. One can't just do it alone, either. I think it needs to be institutionalized. It needs to be part of a community ethos. I think in fact it needs to be part of an entire foreign policy.

For Butler (2004, 28–29), therefore, the recognition of vulnerability can lead to more adequate, namely peaceful, responses to different forms of violence and counter the tendency to react violence with more violence. Alyson Cole (2016, 262) has noted that Butler's view on vulnerability paves the way to "nonviolent interaffectivity". Vulnerability does not in this view imply weakness or inferiority, but rather, it is a human condition, "a basic kind of openness to being affected and affecting in both positive and negative ways, which can take diverse forms in different social situations (for example, bodily, psychological, economic, emotional, and legal vulnerabilities)" as Erin Gilson (2011, 310) summarizes Butler's views. Understanding oneself as vulnerable involves an understanding of the self as being shaped through its relationships to others, its world, power and environs. This has, in my view, implications for the everyday as our embodiment and vulnerability are embedded in the everyday, in it its historicity, forms of power, materiality and concreteness.

Vulnerability is not just a way of referring to the capacity to be wounded. It is also a "way of indicating one's dependency on another, a set of institutions, or a circumambient world to be well, to be safe, to be acknowledged", argues Butler (2011, 200) in the spirit of early feminist peace researchers' work. While vulnerable body is often thought to be private and non-political, when in contact with other bodies the politics of vulnerability comes to fore in the form of, first, the unequal distribution of vulnerabilities and second in the form of accountability, recognition and acknowledgement. In other words, how vulnerability is recognized, accounted for and acknowledged are at the core of the political. The book, thereby, focuses on "the political" as an existential relation that we all live on a daily basis, in the everyday, in ways that create, re-produce, transcend and challenge differences, hierarchies and discriminations as well as vulnerabilities between subjectivities and political positions. The book shows that vulnerable bodies are agentic as receptivity, accountability and vulnerability are actually the presuppositions of agency, not its opposite (Cole 2016, 268–271). In short, vulnerability is enabling for our being in the world. Furthermore, vulnerability is a "condition of potential that makes possible other conditions" (Gilson 2011, 310), including peace as a commitment to live with vulnerability and as community ethos.

Although vulnerability can be seen to be among other things a shared human condition, it is lived and experienced in different ways as well as distributed unequally as argued above. The ways in which we live and are affected, can be understood only in light of the particularity of embodied, social and mundane experiences. It is in this way the feminist and the critical theorizing on the body and everyday as well as on vulnerability open up new pathways to re-theorize peacebuilding and peace as something that becomes expressed, takes place, through acts and points of everyday contact between variously situated and variously vulnerable bodies, namely in corporeal events where accountability, recognition and acknowledgement emerge. The pluralistic and critical approaches indicated above are more sensitive to the changing patterns of dynamic of peace than many abstract approaches to it, and thereby it bears some similarities with the "local turn" theorizing. Yet, it differs greatly from the turn as the new agenda is based on a corporeal analysis that brings together affect, emotions and the somatic and provides an understanding of the body as both the subject and object of discourses, materialities, practices and policies of peace.

Summary

The ontology of being vulnerable and connected guides the book towards new ways of thinking and engaging with the world in general, and peace in particular as it indicates a novel ways of thinking about community and the practices of peace. Encountering vulnerability creates moments of accountability, recognition, response-ability and acknowledgement where the peace ethos of the community is created. Being wounded and being susceptible to vulnerability constitute a rupture in the smooth ordering of political space, and hence it is of a vital importance for peace ethos (cf. Butler 2011). The commitment to living with a certain kind of vulnerability to others and susceptibility of being wounded is in this book the "litmus test" of sustainable peace.

As the body is vulnerable to political and social forces and power as well as to its sheer (biological) materiality, this very vulnerability enforces its connectedness to other bodies. We are always in some sense outside ourselves, exploring or navigating our environment, extended. The subject is invariably formed and maintained within social networks of interdependency. In short, existence takes place between bodies: we are not, but our selves happen whenever they come into contact with others (Puumala 2016). As noted earlier, within such a framework, peace is not a property or structure of a given society, but rather something that becomes expressed, takes place, through acts and points of contact between variously situated bodies, namely in corporeal events where accountability, recognition and acknowledgement unavoidably emerge.

The book offers a heightened sensitivity to the fleshy realities of the human body and by doing this it shows how taking the body seriously introduces phenomenological registers that prioritize the relational and vulnerable elements of human being and, thereby, mundane practices of peace. This view indicates that there is no singular ontology of the body; there are rather ontologies. Since ontology is not singular and given, the question arises about which reality to "know" as there are always specific relevant knowledge that emerge for the ontologies. Furthermore, in this view ontologies do not precede or escape politics, but have politics of their own: the question is about a politics of what is the reality that takes shape and that various people come to live with. In short, I seek to demonstrate that ontologies, "bodily ontologies" (cf. Murphy 2012), can be attuned to the realization of mutual dependence and exposure and that those instances are fundamental for the mundane practices of peacebuilding.

References

Autesserre, S. (2014). Going Micro: Emerging and Future Peacekeeping Research. *International Peacekeeping, 21*(4), 492–500.

Bethlehem, L. (2013). *Violent Seductions: The Mneumonics of the Body in Post-apartheid South-Africa*. Accessed 12 Apr 2018. http://www.newschool.edu/nssr/historymatters/papers/LouiseBethlehem.pdf.

Björkdahl, A., & Gusic, I. (2015). "Global" Norms and "Local" Agency: Frictional Peacebuilding in Kosovo. *Journal of International Relations and Development, 18*(2), 265–287.

Björkdahl, A., & Höglund, K. (2013). Precarious Peacebuilding: Friction in Global–Local Encounters. *Peacebuilding, 1*(3), 289–299.

Boulding, E. (1990). *Building a Global Civic Culture: Education for an Interdependent World*. Syracuse, NY: Syracuse University Press.

Boulding, E. (1996). Peace Behaviours in Various Societies. In *From a Culture of Violence to a Culture of Peace* (pp. 31–53). Paris: Unesco.

Brock-Utne, B. (1989). *Feminist Perspectives on Peace and Peace Education*. New York: Pergamon Press.

Butler, J. (2003, May). Interview with Jill Stauffer. *Believer*. Accessed 28 May 2018. http://www.believermag.com/issues/200305/?read=interview_butler.

Butler, J. (2004). *Precarious Life: The Power of Mourning and Violence*. London and New York: Verso.

Butler, J. (2009). *Frames of War*. London and New York: Verso.

Butler, J. (2011). Confessing a Passionate State—Judith Butler im Interview. *Feministische Studien, 2*, 196–205.

Chandler, D. (2010). The Uncritical Critique of "Liberal Peace". *Review of International Studies, 36*(S1), 137–155.

Cole, A. (2016). All of Us Are Vulnerable, but Some Are More Vulnerable Than Others: The Political Ambiguity of Vulnerability Studies, an Ambivalent Critique. *Critical Horizons, 17*(2), 260–277.

Confortini, C. (2006). Violence, and Gender: The Case for a Peace Studies/Feminism Alliance. *Peace & Change, 31*(3), 333–367.

Confortini, C., & Ruane, A. (2013). Sara Ruddick's *Maternal Thinking* as Weaving Epistemology for *Justpeace*. *Journal of International Political Theory, 10*(1), 70–93.

Crossley, N. (1995). Merleau-Ponty, the Elusive Body and Carnal Sociology. *Body and Society, 1*(1), 43–63.

Das, V. (2007). *Life and Words: Violence and the Descent into the Ordinary*. Berkeley: University of California Press.

Dewsbury, J.-D. (2010). Performative, Non-representational and Affect Based Research: Seven Injuctions. In D. DeLyser, S. Herbert, S. Aitken, M. Crang, & L. McDowell (Eds.), *The Sage Handbook of Qualitative Geography* (pp. 31–54). London and New York: Sage.

Enloe, C. (1996). Margins, Silences and Bottom Rungs: How to Overcome the Underestimation of Power in the Study of International Relations. In S. Smith, K. Booth, & M. Zalewski (Eds.), *International Theory: Positivism and Beyond* (pp. 186–202). New York: Cambridge University Press.

Enloe, C. (2006). *The Curious Feminist: Searching for Women in a New Age of Empire*. Berkeley and Los Angeles: University of California Press.

Firchow, P., & Mac Ginty, R. (2017). Measuring Peace: Comparability, Commensurability, and Complementarity: Using Bottom-Up Indicators. *International Studies Review, 19*(1), 6–27.

Galtung, J. (1969). Violence, Peace, and Peace Research. *Journal of Peace Research, 6*(3), 167–191.

Galtung, J. (1985). Twenty-Five Years of Peace Research: Ten Challenges and Some Responses. *Journal of Peace Research, 22*(2), 141–158.

Gardiner, M. (2000). *Critiques of Everyday Life*. New York and London: Routledge.

Gearoid, M., van der Lijn, J., & Verkoren, W. (2013). Peacebuilding Plans and Local Reconfigurations: Frictions Between Imported Processes and Indigenous Practices. *International Peacekeeping, 20*(2), 137–143.

Gilson, E. (2011). Vulnerability, Ignorance, and Oppression. *Hypatia, 26*(2), 308–332.

Heidegger, M. (1996). *Being and Time: A Translation of "Sein und Zeit"* (J. Stambaugh, Trans.). Albany: State University of New York Press.

Hohe, T. (2002). Clash of Paradigms: International Administration and Local Political Legitimacy in East Timor. *Contemporary Southeast Asia, 24*(3), 569–589.

Husserl, E. (1992). *Grundprobleme der Phänomenologie*. Hamburg: Felix Meiner Verlag.

Jenkins, T., & Reardon, B. (2007). Gender and Peace: Towards a Gender-Inclusive and Holistic Perspective. In C. Webel & J. Galtung (Eds.), *Handbook of Peace and Conflict Studies* (pp. 209–231). New York: Routledge.

Jutila, M., Pehkonen, S., & Väyrynen, T. (2008). Resuscitating a Discipline: An Agenda for Critical Peace Research. *Millennium: Journal of International Studies, 36*(3), 623–640.

Leonardsson, H. (2015). The "Local Turn" in Peacebuilding: A Literature Review of Effective and Emancipatory Local Peace Building. *Third World Quarterly, 36*(5), 825–839.

Mac Ginty, R. (2008). Indigenous Peace-Making Versus the Liberal Peace. *Cooperation and Conflict, 43*(2), 139–163.

Mac Ginty, R. (2011). *International Peacebuilding and Local Resistance*. Basingstoke: Palgrave.

Mac Ginty, R. (2014). Everyday Peace: Bottom-Up and Local Agency in Conflict-Affected Societies. *Security Dialogue, 45*(6), 548–564.

Mac Ginty, R. (2015). Where Is the Local? Critical Localism and Peacebuilding. *Third World Quarterly, 36*(5), 840–856.
Mac Ginty, R., & Richmond, O. (2013). The Local Turn in Peace Building: A Critical Agenda for Peace. *Third World Quarterly, 34*(5), 763–783.
Merleau-Ponty, M. (1962). *Phenomenology of Perception*. London: Routledge and Kegan Paul.
Millar, G. (2014). *An Ethnographic Approach to Peacebuilding: Understanding Local Experiences in Transitional States*. London: Routledge.
Murphy, A. (2012). *Violence and the Philosophical Imaginary*. Albany, NY: SUNY Press.
Paffenholz, T. (2014). International Peacebuilding Goes Local: Analysing Lederach's Conflict Transformation Theory and Its Ambivalent Encounter with 20 Years of Practice. *Peacebuilding, 2*(1), 11–27.
Paffenholz, T. (2015). Unpacking the Local Turn in Peacebuilding: A Critical Assessment Towards an Agenda for Future Research. *Third World Quarterly, 36*(5), 857–874.
Pieterse, J. N. (2010). *Development Theory* (2nd ed.). London, Thousand Oaks, New Delhi, Singapore: Sage.
Puumala, E. (2016). *Assylum Seekers, Sovereignty and the Senses of the International*. London and New York: Routledge.
Randazzo, E. (2016). The Paradoxes of the "Everyday": Scrutinising the Local Turn in Peace Building. *Third World Quarterly, 37*(8), 1351–1370.
Richmond, O. (2001). A Genealogy of Peacemaking: The Creation and Re-creation of Order. *Alternatives, 26*(3), 317–348.
Richmond, O. (2009a). A Post-liberal Peace: Eirenims and Everyday. *Review of International Studies, 35*(3), 557–580.
Richmond, O. (2009b). The Romanticisation of the Local: Welfare, Culture and Peacebuilding. *International Spectator, 44*(1), 149–169.
Richmond, O. (2010). A Genealogy of Peace and Conflict Theory. In O. Richmond (Ed.), *Palgrave Advances in Peacebuilding: Critical Developments and Approaches* (pp. 14–38). London: Palgrave Macmillan.
Richmond, O. (2012). Beyond Local: Ownership and Participation in the Architecture of International Peacebuilding. *Ethnopolitics, 11*(4), 354–375.
Richmond, O., & Mitchell, A. (2011). *Hybrid Forms of Peace: From Everyday Agency to Post-Liberalism*. New York: Palgrave.
Robinson, F. (2011). Stop Talking and Listen: Discourse Ethics and Feminist Care Ethics in International Political Theory. *Millennium: Journal of International Studies, 39*(3), 845–860.
Ruddick, S. (1990). The Rationality of Care. In J. B. Elshtain & S. Tobias (Eds.), *Women, Militarism, and War: Essays in History, Politics, and Social Theory* (pp. 229–254). Lanham, MD: Rowman & Littlefield.
Ruddick, S. (1993). New Feminist Work on Knowledge, Reason and Objectivity. *Hypatia, 8*(4), 140–149.

Ruddick, S. (1995). *Maternal Thinking: Toward a Politics of Peace*. Boston, MA: Beacon Press.

Ruddick, S. (2001). Making Connections Between Parenting and Peace. *Journal of the Association for Research on Mothering, 3*(2), 7–20.

Schutz, A. (1967). *The Phenomenology of the Social World*. Evanston, IL: Northwestern University Press.

Shinko, R. (2008). Agonistic Peace: A Postmodern Reading. *Millennioum: Journal of International Studies, 36*(3), 473–491.

Spivak, G. (1994). Can the Subaltern Speak? In C. Nelson & L. Grossberg (Eds.), *Marxism and the Interpretation of Culture* (pp. 271–313). Chicago: University of Illinois.

Tickner, A. (1992). *Gender in International Relations: Feminist Perspectives on Achieving Global Security*. New York: Columbia University Press.

Vaittinen, T. (2017). *The Global Biopolitical Economy of Needs: Transnational Entanglements Between Ageing Finland and the Global Nurses Reserve of the Philippines*. Tampere: Tampere Peace Research Institute.

Vannini, P. (2015). Non-representational Ethnography: New Ways of Animating Lifeworlds. *Cultural Geographies, 22*(2), 317–327.

Väyrynen, T. (2001). *Culture and International Conflict Resolution*. Manchester and New York: Manchester University Press.

Väyrynen, T. (2005). A Shared Understanding: Gadamer and International Conflict Resolution. *Journal of Peace Research, 42*(3), 349–357.

Väyrynen, T., & Puumala, E. (2015). Bodies of War, the Past Continuous, and (Ar)Rhytmic Experiences. *Alternatives: Global, Local, Political, 40*(4), 237–250.

Väyrynen, T., Puumala, E., Pehkonen, S., Kynsilehto, A., & Vaittinen, T. (2017). *Choreographies of Resistance: Mobile Bodies and Relational Politics*. London and New York: Rowman and Littlefield.

Wallensteen, P. (2011). *Peace Research: Theory and Practice*. New York: Routledge.

CHAPTER 3

Relational and Connecting Body on the Home Front

Abstract The chapter elaborates how people experience war, conflict and peacebuilding as felt and corporeal. It highlights that focusing on the relational and sensing body it is possible to study war and peacebuilding in ways that challenge the prevailing and normalized assumptions about conflict and peacebuilding agency, namely the belief that agency resides solely either in institutionalized politics or is constituted through overt violence. The chapter derives from the narratives of two eldery women who have experienced the Second World War and it underscores the line where the body intersects with politics. As the line is under constant negotiation, it opens up also alternative political horizons for peacebuilding.

Keywords War experience · Agency · Home front

People experience war, conflict, peacebuilding and peace as felt and corporeal. In order to examine these felt and corporeal qualities of war and peacebuilding, this chapter is written around war events and stories of these events, around embodied stories of sensuous and relational human beings whose choreographies form rich entanglements between, for example, war front and home front, citizen and alien, friend and enemy, war and peace. In the events examined in the chapter, the body is revealed to be a "place and time holder" that transfers spatial

T. Väyrynen, *Corporeal Peacebuilding*, Rethinking Peace and Conflict Studies, https://doi.org/10.1007/978-3-319-97259-6_3

and temporal boundaries across space and time. The chapter is founded on Christine Sylvester's (2013) theorizing on war as a socio-political institution that begins and ends with bodies. Similarly, peacebuilding is seen here as a process that begins and ends with bodies as bodies are affected by it and feel it. By focusing on the body, and particularly on bodies coming together, the chapter urges the study of the war and peacebuilding in ways that challenge the prevailing and normalized assumptions about conflict and peacebuilding agency, namely the belief that agency resides solely either in institutionalized politics or is constituted through overt violence. With the recollections of two war stories conducted by Eeva Puumala (Ada's story) and myself (Bertha's story), the chapter makes temporal transitions and travels back in time, yet it never truly leaves the present (Väyrynen and Puumala 2015). The stories open intriguing spatio-temporal spectres, a window of time to a history and present that is corporeally intertwined with post-war peacebuilding, reconstruction and nation-building.

The diffractive reading of the body in the chapter relies on a relationship between space, time and the body that is not pre-given, but rather negotiated in the heat of the moment, and furthermore, that is sensuous. In short, the chapter seeks to show the way in which a focus on the body as a lived entity and its vivid set of experiences can illustrate the impossibility to unambiguously distinguish between the personal, national and international, past and present, war and peace and even the self and the other. By doing this, the chapter shows how the seemingly spatially "local body" is a nexus of a variety of forces, and how agency emerges from the body.

The chapter relies on a notion of the body put forward the feminist peace researchers discussed in Chapter 2. Their view is complemented with Jean-Luc Nancy's (2000, 2008) theorizing on body as for him the body refers to the necessarily relational taking place of existence. Since relation always concerns more than one body, it opens the question of the political taking place between people. Furthermore, the notion of the body I suggest in this chapter rejects the Cartesian dualism, the separation between mind and body. Instead, I discuss in the chapter the way in which the body and sense are inseparable together with the political implications that the inseparability brings to bear for Peace and Conflict Studies. The type of Nancian reading of sense which I use conceives sense to be inextricable from the body as it designates the following things: the faculties by which the mind receives information about the body or the external world; meaning; reason or purpose; direction and understanding. The view in the chapter on experience underscores the inseparability of the body and experience as experience indicates immediate sensory presence and corporeal plenitude. Accordingly, the adopted notion of experience is closely related to the sensuous effect, senses given to things, movement, the pulse and rhythms of life and the way in which the body senses the pulse of life and rhythmically navigates in space at the very second it experiences the world or certain elements of the world (Nancy 2008, 23–25). A focus on the body, sense and experience allows me to examine how the past conflict flows through and penetrates the present peacebuilding, both in terms of personal and familial histories and post-conflict national identity projects.

Bodies in Transition

Ada is now almost 90 years old and she reminisces her experience of the Second World War. She tells that she was the second youngest of nine siblings. She recounts that nobody from the family went to the Winter War between Finland and the Soviet Union that was fought from 1939 till March 1940. The truce that ended the Winter War did not signify real peace: Europe was still at war and the country remained in a state of exception. Indeed, for Finland the Winter War was merely the first part of the country's experience of the Second World War. In 1941 the Continuation War broke out. At that time Ada was 14 years old and she remembers how her eldest brother Juha was mobilized then, although according to Ada, he had never had any proper military training. Juha served as a pioneer, building bridges and doing that kind of things. Also her brother Kalevi, at the time only seventeen years old, volunteered and was sent to the front. Kalevi was sent to Karelia, the Southeastern border, where he first dug trenches and built dugouts. But as the war went on, Kalevi came of age and was sent to fight in the frontlines. Even though Ada's family lived far removed from the actual battlefields, in the Westernmost Finland, in a small, rural village, the war had an impact also on everyday life and its mundane practices on the home front.

War is about the body and it concerns the body and the body's relationality. Ada's story is not that of a soldier, but one told by a person who at the time was an adolescent girl living far removed from the actual battlefields. Her story does not solely focus on war as a politics of injury, that is, according to Sylvester (2013, 5), the actual practice of war.

Rather, it brings forth the understanding how the experiences of war and peacebuilding are truly corporeal. The story is a narrative by a witness of wartime and of a family's life after the war, but between the lines it is as much a story how some lives are wasted for the benefit of a myth called "unitary nation" and "community" and how the war touches the most intimate relations of the bodily existence (Butalia 2002, 149). It is a story that persuades, in my view, to understand that our existence is in multiple ways tied to others; other bodies, other spaces and temporalities. Moreover, Ada's account enables us to grasp that the connection is so tight that in the end the whole notion of the singular autonomous subject becomes redundant.

The beginning of a war and violence involves a massive mobilization of bodies, especially corps and troops that are sent to the front as Ada's story indicates. Bodies are transferred often also involuntarily. Another woman who had experienced the war, Bertha, reminisced forty years after the war how it meant for her forced migration from the occupied territories to Southern Finland, then back to the territories and finally to Western Finland. She was hence forced to leave twice. Her and her children's bodies were put into crowded and dark cattle wagons to be transported away from the battle-lines. The odors and voices were something they had not experienced before as they had lived ordinary family life in the Karelian region which was later to become a site of intensive warfare. Bertha told how the children were crying and cows who were loaded to same wagons were shouting in a sheer horror. Bertha's husband was left behind because he was needed for organizing the transference of the local population away from the battle-lines. These events tore families apart as some bodies were forced—or volunteer—to leave and others stay put.

In that way, as Jenny Edkins (2006, 109) claims, war betrayed their faith in the imagined personal and national order. In such a psychoanalytically informed reading of war the state is the organization that sends people to their deaths by ordering them to combat in its name, for its cause and interests. In Bertha's narrative, the state put her family life to disarray. In the story Ada tells, on the other hand, there is no sense of betrayal or resentment present as she claims that leaving for war was a natural thing to do. In Ada's story, the national order seems to coincide with her personal order and its sense of normality. The presented view coincides with Christine Sylvester's (2013, 4) claim that conflict is a social institution that "emerges over time and dominates alternative ways of living to such a degree that they seem normal and natural, or at least unavoidable". What may seem to be exceptional at one point of history, at another turns into normality, and the line that separates these two windows of time is by no means clear-cut. It can be grasped from Ada's and Bertha's accounts that war takes a concrete hold on the human body and its ways of behaviour also beyond the body of the soldier. As Ada in her interview recounts, the impact of war on the body was enormous also at the home front, let alone in the occupied territories and their liminal spatio-temporal existence indicated in Bertha's story.

During the Continuation War Finland was an ally of Germany and the country served as a station place for some 40,000 German soldiers. Ada reminisced how the small city close to her village was full of German soldiers. The locals tried to avoid them, but some women had affairs with soldiers, and children were born. Ada vividly remembers the international presence of German soldiers in the provincial town. In her story, the notable thing was not the pervasive military presence as such but the open sexual interest that the soldiers showed towards Finnish women. She claims that she was too young to attract any insinuations from the side of the Germans, but certainly she had keenly observed the encounters between the Germans and Finns.

Multiple temporal and spatial transitions characterized embodied experiences of war and violence. Bertha's memory leaped towards the Soviet prisoners of war who were kept in the camps in Finland in inhuman conditions. Their hands had reached towards her and begged for bread. She remembered their strange language and alien manners as they expressed overwhelming gratitude towards the Finns who gave them food. Any communication between them and the Finnish citizens living near the camps were forbidden, but some locals disobeyed the orders. For Ada and Bertha, hence, the presence of alien soldiers in home front was an experience that altered or tweaked their sense of the rhythm of her ordinary life. It constituted an event of dislocation, a time-space in which everydayness and familiarity intertwine with and become for a while inseparable from war and the alien.

Power is invested in and forced upon the body during war and also in the post-war period through processes of subjectivation as these stories demonstrate (cf. Åhäll 2012; Zalewski 2000). The corporeal choreographies that different bodies are able to—or forced to—undertake and create during wartime and peacebuilding are complex. In Ada's and Bertha's stories macro-politics intersects and overlaps with micro-politics and mundane to constitute experiences of war, and later peacebuilding and peace. It follows that in Ada's and Bertha's corporeal experiences time cannot be located within a chronological and causal frame. Rather, their bodies evoke a sense of non-linear and interrupted time, where both the past and present intertwine. The excitement in Ada's voice when she reminisces the presence of the German soldiers indicates that her body carries within it sensuous memories of a sensuous presence. As indicated earlier, it is possible to claim that people experience political processes as felt, corporeal memories as the body is exposed to those processes and as it simultaneously withdraws from being completely captured by those processes (Bakare-Yusuf 1997; Dossa 2003). Ada's corporeal memory consists of, for example, the silenced stories of Finnish women's illegitimate children with German soldiers that the post-war nationalistic history writing in Finland has sought to suppress (see Chapter 5 and Väyrynen 2014). Furthermore, Ada's and Bertha's narratives are testimonies of women's non-recognized agency and authority over their lives during wartime, even if that agency meant that they in the official narratives betrayed the nation by associating themselves with the German soldiers or Soviet prisoners of war.

Through the interpretive lens of non-linear time, it is possible to claim that Ada's and Bertha's stories also points towards a particular spatiality. Here it is important to note that space does not indicate a geographically delineable area or a place marked by a specific purpose but, rather, space is an interval that is inherently open for different interpretations and in which no direction is yet more prominent than another one. This interpretation derives from Jean-Luc Nancy (2008, 27, 67) for whom the world is a tension of place, where bodies are not in space, but where space is in bodies. The suggested take on space turns the body into a "place-holder" (Nancy 2008; Puumala and Pehkonen 2010, 58) that pulsates with war and peacebuilding as well as refuses any attempts to be exhaustively written and known.

Temporal Transitions from War to Peace

Bodies resonate with the rythms of war and peacebulding. There are two rhythms, for example, present in Ada's and Bertha's stories. On the one hand, they describe the rhythm between the body and practices of governance that seek to constrain people's behaviour, and on the other, they speak of the always present possibility of the change of rhythm that disturbs those practices of governance (Pehkonen 2016; Puumala and Pehkonen 2010). To underscore the inseparability of the two, the notion of (ar)rythmic experience captures, in my view, the dynamics of peacebuilding.

The female body in war is not necessarily that of a weak victim, but rather a strong working body that fully contributes to the rigid choreography of war. After the men had left, there was a lot of work to do as indicated in Ada's and Bertha's stories. The general view was that every "proper human being" was taken to the front; there were only the old and the frail left—and children, adolescents and women. During the war women continued the hard manual labor that the men had done before leaving. Due to the hard nature of certain tasks and the shortage of labor created by mobilization, some of the most straining and labor intensive chores were jointly organized. Women were working in the fields and factories carrying out heavy manual labour and taking care of the basic functions of the society. Women took the responsibilities that had traditionally belonged to men and run the society in ways that gave birth to new forms of gendered agency. The hard labour continued during the reconstruction, hence, the female body and its agentive capacity remained at the core of the peacebuilding nation.

As Cynthia Enloe (2010) has suggested, the gender politics of war shifts the balance of opportunities and dangers for men and women. With Ada's and Bertha's mundane experiences it is possible to further underscore that the opportunities and dangers that come to exist during violence and conflicts are by no means self-evident and pre-existing. Ada brings forth an event that created a sense of risk. Despite the fact that was strictly prohibited to dance during the wartime, the official orders were defied and dances were organized in barns so as to thank the people who had participated in harvest or planting. Some people found their spouses at those dances when the soldiers on leave arrived to them. Many affairs got started and many children were conceived at these social gatherings. Hence, every now and then the body escaped the touch of war and took risks by doing so. The bodily choreographies that were in part generated by the experience of war became disruptive moves that were fluid and ambiguous as the bodies can enjoy sensuous pleasures. In these events, those sensuous pleasures withdrew the body from the rigid governmental choreography of war and interrupted its functioning in the everyday. These events took shape during the conflict and violence and as such testify of the body's capacity to change its rhythm and speak of experience as the rhythm of bodies.

Often war stories tell us about unimaginable things becoming part of everyday life, but Bertha's account also points towards the imaginative that transforms the presence of the unimaginable in everyday life. While the practices of governance tend to instrumentalize the body particularly during war, Bertha's action escapes the instrumentalization and represents a gestural act of extension from one body to another, of being taking place in an interval. The notion of "interval" (Nancy 2008, 81–97) here designates a space that is not closed with meanings, but rather a space in which two bodies create a shared meaning together and thereby, cross the boundaries and differences created by the nation and its production of enemies. Her response to the vulnerable Soviet prisoners of war certainly represents a negotiation of risk and as such her act tells of both acknowledgement across differences as well as of resistance and empowerment, which yet gave rise to fear and shame. The event of offering food to the prisoners can be read as a commentary to the institution of war and its structures of domination that seek to establish clear differences and categories between us and them, the self and the other, war and peace, the citizen and the alien. Ultimately, Bertha resists the dominant truth claims that sought to produce seemingly clear-cut boundaries between bodies. Her action emerged in the heat of the moment and constituted an immediate and accurate response to the pain and presence of the vulnerable other (cf. Butler 2004, 24).

As the war went on bodies begun to return from the frontlines. The coffins were sent to the train station from where the soldiers had left and from there their families could pick up the coffins by horse carriage. In some coffins the lids were nailed tightly and there was a note on top with a strict order not to open it. In those cases the families could not see the returned body ravaged by war practices. Many families received a long feared visitor of letter telling that a family member had lost his life or seriously wounded in the frontline. People joined forces to guarantee a worthy funeral for the dead soldier. In June 1944 massive battles took place in the Karelian Isthmus and the Southeastern parts of Finland, during which Finnish forces suffered severe losses. That was when Bertha was forced to leave her home the second time.

During the war funerals are important events that bring the community together. While grief is often thought of as privatizing and depoliticizing, it is, however, a source of a sense of community that enables alternative agencies to be born and relations to be negotiated. Pain, sorrow and mourning integrate individual and collective conditions. The affective dynamics of mundane, yet collective rituals, such as funerals, are seldom recognized in the mainstream study of war and peacebuilding. In funerals the political element of pain emerges when the mourners make the vulnerability of the existence visible, claim their truth and create an "affective enclave" (Seremitakis 1991, 5). In such enclaves emotions provoke responses and reciprocation from others. They build affective exchange and consensus as they are based on emotional and communicative interchange and choreography. Ultimately, in these enclaves individual emotional communication intertwines with larger social and political structures. The mourning subject is in conflict with the institution of war that has sent the person's loved one to his or her death. However heroic the individual death from the nationalistic perspective may be, the pain of the mourners cannot be overridden. The dead heroic body and the living mourning body are inseparable and they point towards the inseparability of the collective from the personal, mundane and corporeal as well as war from peacebuilding. Thus, the relationship between the body and the collective is far from clear-cut; the governmental rhythm between the two is under constant risk of disruption. In funerals that take place during violence and war, mourning bodies became often disruptive and disjunctive spaces with their distinctive choreographies that created alternative social structures.

Besides the fact that wars result in dead bodies, also a massive number of bodies are deeply altered in war and further yet, as Elaine Scarry (1985, 113) reminds us, those alterations are carried in the body after the war to peacebuilding, reconstruction and reconciliation. Such corporeal alterations are most visible and recognisable in the wounded and mutilated bodies of war veterans. Accordingly, as Marita Sturken (1991, 132) writes, "[the war] veteran's bodies—dressed in fatigues, scarred and disabled, contaminated by toxins—refuse to let historical narratives of completion stand" (cf. Väyrynen 2012). These bodies constitute an unstable signifier in the system of war as they can both be encoded to represent heroisms and, at the same time, through their disfiguration constitute an vulnerable abject in the time of peace as it will be demonstrated in the following chapter (Chapter 4).

One day Ada's family had a visitor who brought the message about her brother going missing in action. Although the family never talked about him being missing and just waited for him to return, Ada developed her own story what happened to her bother: she imagined him being drowned during the retreat of the Finnish army under the major Soviet attack. This part of Ada's account illustrates perhaps most strikingly the centrality of the body in her story. It is not the wounded body, but the missing body whose non-presence and non-materiality escapes the narrations of war and haunts her experience of peacebuilding and peace (Edkins 2011; Clair 1997; Das 1998; Ross 2001). The missing can neither represent the heroic body of a soldier killed in action, nor can the loss be shared and communicated within the local community. No affective enclave can be formed around a missing body because, as Ada points out, it was not certain that there was a body at all. She describes that the family experienced that Kalevi was "nowhere", neither dead nor alive. Kalevi's body turns into an immaterial or impossible body that has vanished in Ada's retrospective interpretation into the cool depths of water. There is no way to physically reach towards him; a sense of the incomplete and inexplicable will always accompany Kalevi's life—and presumable death. The event is representative of a situation when the missing body becomes incorporated into one's corporeal memoryscape in a particular way since the sense of loss escapes verbal representations. With disappearance further "layers of silences" (Butalia 1998, 127) emerge in Ada's story, as she fails to capture the missing and the unspeakable experience that through her behavior and gait takes shape as a corporeal choreography of pain. That choreography provides no sense of closure and constitutes an incomplete experience where decades make the body to suffer over and over again.

(Ar)rhythmic Experiences

The cost of peace is often high both in human and financial terms. In Finland, over 63,000 men had been lost in war, Finland had ceded large parts of its territory and was to pay 300 million dollars as reparations to the Soviet Union. That meant hard work also in relocating the people, such as Bertha's family, from the ceded territories. After the war, the atmosphere in the society had changed. Men returned from the front, but in some cases the women had new spouses and children with them. Men who returned had suffered mental and physical wounds (Kivimäki 2013). And not all Finnish men had followed the orders to go to conscriptions so they lived in the forests during the war. Some men, as Bertha's husband, were not fit enough to go to war and were deemed eligible onto to the auxiliary tasks. Hence, there were tensions between those who went to war and those who deserted, who were discharged or whom were too young to join the army.

War accentuates certain myths about male and female identities through which the future of the nation is inscribed in the life-giving capacity of the body. That line of thought easily prescribes women the role of waiting for men to return dead or alive. However, Ada's account of women's role as life-givers actually enables, rather than denies, agency. In her story some women refused to wait for and "save themselves" for the Finnish men, even their husbands, during the war. The story emphasizes the body, to use Sylvester's (2013, 72) words, as "a vessel of pleasure and choice", not as passive victim of war. In a similar vein, the body of a deserter and discharged breaks the myth of a sacrificial male body that finds its place in the institution of war with ease (cf. Hutchings 2008). The question of deserters and discharged brings again forth an understanding of the impossibility of total disciplining of the body. These bodies betray and interrupt the rigid choreography of war that lies at the heart of the national body politics.

Ada refers to the post-war collective silences in her account that were related to the deserting and unfaithful bodies. Both the male deserter's body and the female body that refuses to abstain from sexual relations—and even the missing or discharged body to a certain extent—resist the national grand narrative of joint war effort and the unity of the national body politic in the face of violence. The existence of those silences points towards the necessary limits for the formation of the future collective identity project at the societal level, but on the other hand, the story signals the knowledge of and passing on of those narratives that could not be accommodated in the official national history writing (Fierke 2008; Duara 1995; Hobsbawm 1992; Olick and Robbins 1998; Nora 1989). Thus, the experience of war that can be read from individual stories is more fine-tuned and processual than the official narrative of the Finnish national identity project during and after the war would suggest.

These stories open specific windows of the temporal transitions between the past and the present and extend corporeally yet further to the future. Marianne Hirsch's (2008) notion of "postmemory" captures the temporal transition as it refers to the second-generation traumatic and embodied knowledge. Postmemory is generated when the effects of past events continue in the present. It is corporeal in the sense that it is transmitted through the language of the body. Because it is corporeally transferred, postmemory differs radically from national official memory that is passed on largely through the symbolic system of language. Furthermore, unlike postmemory, post-war national memory seeks closures that allow the creation of the image of a unitary national self. As Ada's and Bertha's stories and the multiple silences that exist within them indicate, there is no signifier that could fully represent the nation during peacebuilding and create an unitary national self. In accordance with that thought, the nation remains distorted in the sense that the internal coherence that is assumed to characterize the peacebuilding community within proves itself to be a false promise (cf. Žižek 1997).

Summary

The body in war and peacebuilding is a mobile body both in terms of its physical, and even sensuous, movement and its resistance to comprehensive and essentializing differences and categorizations. Ada's and Bertha's accounts in this chapter expose the intertwining of the practices of discipline and the inherent creativity and reaching that is characteristic of the body. They illustrate that the line where the body intersects with politics is under constant negotiation, which can open alternative political horizons for peacebuilding. Even though some movements during conflicts are strategically pre-written, everyday life entails equally corporeal, and yet non-linear relationalities between people, who extended and reached out towards one another. Through the rich variety of corporeal choreographies that people undertake during violent political conflicts, the stories in this chapter shows poignantly that risk and control are negotiated both in the production of an alternative space for action and through physical relationships.

As the stories have illustrated, senses and experiences of war and conflict are physical, emotional and sensuous all at once. The institution of war penetrates the body thoroughly, but in unanticipated ways the body can refuse to conform to the limitations that are imposed through the practices of governance. Bodies' (ar)rhytmic performances are a part of the institution of war and through them the lines between violence and peacebuilding, the past and present and the self and other become ambiguous. As Ada's and Bertha's accounts illustrate experience neither produces nor is produced by a stable or substantial identity that would speak from a particular standpoint. The experience of war and its aftermath, namely peacebuilding, remain shifting and open, but even so they do leave both physical and emotional traces on the body. This chapter has demonstrated that engaging with the body and the sensuous experience of existence during wartime and peacebuilding, it is possible to shed light on the multiplicity of relationships and transitions between space, time and the body, whether they follow the prewritten choreography of political life or disrupt it by withdrawing from the practices of governance.

References

Åhäll, L. (2012). The Writing of Heroines: Motherhood and Female Agency in Political Violence. *Security Dialogue, 43*(4), 287–303.

Bakare-Yusuf, B. (1997). The Economy of Violence: Black Bodies and the Unspeakable Terror. In R. Lențin (Ed.), *Gender and Catastrophe* (pp. 171–185). London and New York: Zed Books.

Butalia, U. (1998). *The Other Side of Silence: Voices from the Partition of India*. New Delhi: Punguin Books.

Butalia, U. (2002). A Necessary Journey: A Story of Friendship and Reconciliation. *Alternatives, 27*(2), 147–164.

Butler, J. (2004). *Precarious Life: The Powers of Mourning and Violence*. London: Verso.

Clair, R. (1997). Organizing Silence: Silence as Voice and Voice as Silence in the Narrative Exploration of the Treaty of New Echota. *Western Journal of Communication, 61*(3), 315–337.

Culbertson, R. (1995). Embodied Memory, Transcendence, and Telling: Recounting Trauma, Re-establishing the Self. *New Literary History, 26*(1), 169–195.

Das, V. (1998). Language and Body: Transactions in the Construction of Pain. In A. Kleinman, V. Das, & M. Lock (Eds.), *Social Suffering* (pp. 67–92). New Delhi: Oxford University Press.

Dossa, P. (2003). The Body Remembers: A Migratory Tale of Social Suffering and Witnessing. *International Journal of Mental Health, 3*(3), 50–73.

Duara, P. (1995). *Rescuing History from the Nation: Questioning Narratives of Modern China*. Chicago: University of Chicago Press.

Edkins, J. (2006). Remembering Rationality: Trauma Time and Politics. In D. Bell (Ed.), *Memory, Trauma and World Politics* (pp. 99–115). London: Palgrave Macmillan.

Edkins, J. (2011). *Missing: Persons and Politics*. Ithaca: Cornell University Press.

Enloe, C. (2010). *Making Feminist Sense of the Iraq War*. Los Angeles: University of California Press.

Fierke, K. (2008). Whereof We Can Speak, Thereof We Must Not Be Silent: Trauma, Political Solipsism and War. *Review of International Studies, 30*(4), 471–491.

Hirsch, M. (2008). The Generation of Postmemory. *Poetics Today, 29*(1), 103–128.

Hobsbawm, E. (1992). Ethnicity and Nationalism in Europe Today. *Anthropology Today, 8*(1), 3–8.

Hutchings, K. (2008). Making Sense of Masculinity and War. *Men and Masculinities, 10*(4), 389–404.

Kivimäki, V. (2013). *Battled Nerves: Finnish Soldiers' War Experience, Trauma, and Military Psychiatry, 1941–44*. Turku: Åbo Akademi University.

Nancy, J.-L. (2000). *Being Singular Plural*. Stanford: Stanford University Press.

Nancy, J.-L. (2008). *Corpus*. New York: Fordham University Press.

Nora, P. (1989). Between Memory and History: Les Lieux de Mémoire. *Representations, 26*(2), 7–24.

Olick, J., & Robbins, J. (1998). Social Memory Studies: From "Collective Memory" to the Historical Sociology of Mnemonic Practices. *Annual Review of Sociology, 24*, 105–140.

Pehkonen, S. (2016). Choreographing the Performer-Audience Interaction. *Journal of Contemporary Ethnography, 43*(3), 306–330.

Puumala, E., & Pehkonen, S. (2010). Corporeal Choreographies: Failed Asylum-Seekers Moving from Body Politics to Bodyspaces. *International Political Sociology, 4*(1), 50–65.

Ross, F. (2001). Speech and Silence: Women's Testimony in the First Five Weeks of Public Hearings of the South African Truth and Reconciliation Commission. In V. Das, A. Kleinman, M. Lock, M. Ramphele, & P. Reynolds (Eds.), *Remaking a World Violence, Social Suffering, and Recovery* (pp. 250–280). Berkeley: University of California Press.

Scarry, E. (1985). *The Body in Pain: The Making and Unmaking of the World*. Oxford: Oxford University Press.

Seremitakis, N. (1991). *Last Word: Women, Death, and Divination in Inner Mani*. Chicago: Chicago University Press.

Sturken, M. (1991). The Wall, the Screen, and the Image: The Vietnam Veterans Memorial. *Representations, 35*, 118–142.

Sylvester, C. (2013). *War as Experience: Contributions from International Relations and Feminist Analysis*. New York: Routledge.

Väyrynen, T. (2012). Keeping the Trauma of War Open in the Male Body—Resisting the Hegemonic Forms of Masculinity and Finnish National Identity. *Journal of Gender Studies, 22*(1), 137–151.

Väyrynen, T. (2014). Muted National Memory: When the "Hitler's Brides" Speak the Truth. *International Feminist Journal of Politics, 16*(2), 218–235.

Väyrynen, T., & Puumala, E. (2015). Bodies of War, the Past Continuous, and (Ar)rhytmic Experiences. *Alternatives: Local, Global, Political, 40*(4), 237–250.

Zalewski, M. (2000). *Feminism After Postmodernism: Theorising Through Practice*. London: Routledge.

Žižek, S. (1997). *The Sublime Object of Ideology*. London: Verso.

CHAPTER 4

Returning Body: Transferring Violence Across Time and Space

Abstract Bodies that are touched by war and violence transfer war across spatial and temporal boundaries to peacebuilding and peace. It follows that post-conflict space is filled with bodies that are touched by violence. This chapter engages with witnessing existence and the narratives of war veterans and demonstrates how the corporeal and intersubjective relations of these bodies transfer the touch of violence to other bodies. Starting from Karen Barad's new materialism, the chapter argues that corporeal matter has agency to bring distant warfare home as it becomes felt in bodies dispersed in time and space. This observation is vital for peacebuilding since the boundary between "us" and "them" becomes reconfigured when the witnessing body circulates the touch of war across time and space.

Keywords Mechanisms of conflict resolution · Witness
New materialism · War veteran

Bodies that are touched by war transfer war across spatial and temporal boundaries. As Ada reminisces in the previous chapter (Chapter 3), the return of soldiers to home front is not an unambiguous event: returning bodies create differences within the community as some bodies are uplifted to present the nation and others unacknowledged, even despised. The injury on the body war and violence create is not just physical one since it always touches the other layers of being a human as

T. Väyrynen, *Corporeal Peacebuilding*, Rethinking Peace and Conflict Studies, https://doi.org/10.1007/978-3-319-97259-6_4

body and mind are inseparable. A fictional narrative of a Finnish female war veteran Anu is illustrative of this body-mind connection which is repeated in many war veterans' narratives. She tells about her life after her return back from the ISAF mission in Afghanistan to Finland.

> I seal off my mind from things that remind me of having been there. Instead, I try to live my life fully now here at home. [...] It is difficult to admit that one is somewhat mentally stuck. And there is no obvious reason for that. How can one feel this low and depressed, although everything went eventually quite well and nothing serious happened to me in Afghanistan?

Anu suffers from sleep disorder and a sense of being alienated from her friends. She is keen on returning to something which would create normalcy and the familiar rhythm of everyday life, i.e. to mundane, and yet she has a feeling that things are not quite right. Despite her attempts to cling to the ordinary and familiar, things appear for her in an eery way unfamiliar.

This chapter takes a leap in time by discussing recent warfare and war bodies. It is again written around war narratives. This time the narratives were told by Finnish soldiers that took part in the NATO-led ISAF mission in Afganistan, 2001–2014. The material is collected from the variety of interviews the war veterans gave to the Finnish media after their return home. Many of the interviews follow a similar pattern, namely, the veterans describe their experiences of war and the difficulties of returning in a way the fictive figure Anu does in the quote above.

Whereas the previous chapter relied on the "relational" ontology, in this chapter I cultivate new materialist ontology. The chapter examines how bodies are touched by war and violence in a multiple manner, how this touch of violence is transferred across time and space and, furthermore, how the touch makes peacebuilding a long-term affective process. In short, the chapter focuses on how violence is transferred through material bodies. As argued in the previous chapter, war continues and ends in bodies (cf. Sylvester 2011, 1). Disfigured and mutilated bodies transfer war from the battlefields to the home front. For most of the Western population, battlefields are remote and the population encounters war and violence in the form of war veterans' bodies: veterans' mutilated and disfigured bodies frame conflicts for the entire population.

Homi Bhabha's (1992, 146) writing on the transnational histories of migrants, the colonized and political refugees holds relevance to my argument in this chapter. He considers these to be "border and frontier conditions", and the chapter builds on Bhabha's work in the sense that it seeks to demonstrate how bodies touched by warfare, e.g. soldiers, peacekeepers, civilian crises management personnel, aid workers etc., do not just representationally frame the war to larger population, but this framing is corporeal when material bodies bring distant warfare home due to their "freak displacements" (Bhabha 1992, 146). For instance, Anu's body is a reminder of this kind of "freak displacement" as it seems to belong nowhere, neither war nor peace, neither there nor here, neither Afghanistan nor Finland. This chapter, thereby, continues the discussion started in the previous chapter, namely how events, structures of feeling, choreographies and the backgrounds of everyday life (cf. Vannini 2015, 317) open new horizons to the study of war and peacebuilding. I seek to show that when war and peacebuilding efforts take place in distant places and remote locations, the bodies that have experienced war bring violence home. That was the case with Ada's brother who came home from the frontlines when the Second World War ended, and it is the case with, for example, a Western war veteran who has taken part in an international peacebuilding mission.

While uplifted to represent the nation particularly damaged bodies that return are at the same time constant reminders of the trauma of the war and the vulnerability of the human bodily existence (Väyrynen 2014, 2016). Furthermore, as indicated earlier, the veterans' bodies have an ambivalent relationship to the nation since they constitute a form of abject existence (Sturken 1991, 132). They challenge the nation's image as an ultimate provider of human security while, at the same time, heroic narratives can be told through them (Edkins 2003). The figure of the veteran gets an additional affective intensity when the disfigurement takes place at the mental level. By creating "freak displacements" mentally disfigured war veteran's bodies disturb the hegemonic representations of "heroically" disfigured bodies.

Another starting-point for this chapter is Christine Sylvester's (2012) work where she urges us to focus on concrete human bodies when we study war and conflict. Her (Sylvester 2012, 484) work shifts the focus also to agency, and particularly to war agency where the body and its agentic capacity emerge through war and violence. Her research agenda does not take part in the study of the representations of war, i.e. how war is represented in and through verbal and visual rhetoric, but it rather turns the attention to the material forces that constitute war body and war agency. Sylvester's research agenda hence points toward the study of material, performative and relational bodies that are touched by war in multiple and overlapping ways. What is, in my view, left under-theorized by Sylvester is the transference of the touch of war.

In short, in this chapter I seek to outline how feminist theory, such as Sylvester's theorizing on war, and particularly feminist new materialism offer novel ways of thinking how violent conflict touches us and how this touch is transferred across time and space to the spatial-temporal constellations of peacebuilding. The chapter seeks to think through how the bodes that are touched by war touch us who have indirect experiences of it. The ontology of the chapter is that of the new materialist ontology as the chapter engages with the corporeal-material touch of war (cf. French 1994). The chapter provides also a diffractive reading, as indicated in the first chapter, where new insights are built by "attentively and carefully reading for differences that matter in their fine details, together with the recognition that there intrinsic to this analysis is an ethics that is not predicated on externality but rather entanglement" (Barad 2003, 811).

Although the case study example is on Finland again, it has relevance to many of the Western societies that engage in distant warfare through peace enforcement, peacekeeping and military missions abroad. Moreover, the discussion on the touch of war and its transition through peacebuilding to peace holds relevance to any society that has experienced violent political conflict. After introducing the Finnish context and Finland's engagement in distant warfare in the light of war agency, the chapter discusses how war and violence has touched the Finnish population. The Second World War and peacekeeping bodies serve as a background for the reading conducted through the case of the Finnish involvement in the NATO-led ISAF mission. I examine how the distant war and violence comes close to us and how the non-representational transference of them takes place. Overall, the chapter examines how distant violence touches us, how warfare and conflict is brought home and what kind of ethical implications the touch of conflict has.

Heroic and Benign War Bodies

Wars provide often an anchor for the master narratives around which the collective identity is constructed. For example in Finland, the national identity revolves around the Second World War where the nation fought against the Soviet Union. The war is constitutive of the nation's identity in the sense that the nation's present existence is seen to be brought fore by the sacrifices made during the war. Finland is presented in the archival national history writing as having fought alone, heroically and honourably against its mighty enemy, namely the Soviet army, entailing major sacrifices. As indicated in the previous chapter, Finland formed an alliance with the Nazi Germany during the war. More than 200,000 German military men traversed or were stationed in Finland over the three years when Nazi Germany's occupation of Norway brought to the fore the need to transfer troops and munitions through Sweden and Finland. When Germany declared war on the Soviet Union in June 1941, Finland grew closer to Germany and eventually became a co-belligerent with Nazi Germany (Lunde 2011; Kivimäki 2012). The Lapland War in 1944–1945 followed the signing of an armistice agreement with the Allied Powers and it stipulated expulsion of German forces from Finnish territory.

In Finland, as in many other places which have experienced conflict, the sacrificial, strong and unyielding male body has had a charged presence in the post-conflict affective landscape where the nation has sought to re-establish its identity narrative. Heroic Finnish male bodies are seen to have sacrificed themselves, particularly during the winter of 1939–1940 in the "Winter War". During the winter Finnish soldiers fought several months longer than the Soviets assumed although the Soviets employed three times more man power that the Finns. The Second World War was a typical territorial war for Finland: it concerned Finnish territory, it was fought to a large extent close to the Finnish pre-war borders and ground troops were the main form of military power employed. The war meant a great number of battle casualties, i.e. 95,000 deaths which is almost 3% of the population and 195,000 wounded of which 95,000 were permanently disabled. This war touched directly a great number of Finnish families as the narratives of Ada and Bertha in the third chapter demonstrated. Thereby, the Second World War veterans' corporeality, or at least the heroic kind, has had strong presence among the Finnish population. As the mobilization of the troops was very extensive as the Ada's story in the previous chapter (Chapter 3) indicated, most of the Finnish families were indeed touched by the war. As also indicated, many combatants suffered from a variety of disabilities and many families had to deal with disabled war veteran's bodies.

Finnish soldiers' engagement in different UN peacekeeping missions has produced another type of war body. Instead of producing heroic war body, benign male body has emerged. Since the Suez operation in 1956, about 50,000 Finns have voluntarily served in more than 40 operations. These operations have taken place in Middle-East, Africa, Asia and Europe and 39 Finnish citizens have died when serving in peacekeeping missions. In 2015, 4000 peacekeepers got a war veteran's status. Although Finns have wounded and died in peacekeeping, their bodies have not been elevated to a sacrificial position. They are seen to represent peace rather than war, not being targets of violence nor touched by war in a manner the Second World War veterans are seen to be. These bodies have had very limited affective presence among the Finnish population, although they are employed in the national foreign policy doctrine that hails Finland into being as "peacekeeping superpower". In sum, the presence and touch of war body in Finland has changed as the Second World War veterans' bodies are getting less visibility and the peacekeepers' bodies do not occupy any central place in the Finnish affective landscape. Peacekeepers bodies have been seen as benign, almost "feminized" (Enloe 1993; Sanghera et al. 2008), whose war agency is minimal compared to the Second World War bodies and their charged affective agency.

Body Touched by War

Unlike the Finnish peacekeepers and their benign war agency, the war in Afghanistan produced corporeality that has intensive war agency. War veterans bodies form an interesting case study material as they bring distant warfare and battlefields home in multiple, and often forceful, ways. As described above, there is a long tradition of Finnish peacekeeping in international missions abroad, but the case of Finnish involvement in NATO-led ISAF operation in Afghanistan was different: in Afghanistan Finland deployed military units and the mission engaged Finland in distant warfare. Finnish troops were regularly in fire-fight and four persons, two soldiers and two humanitarian aid workers, died during the operation. The justification of the non-NATO country's participation in the NATO operation was seen to be necessary many times during the mission. A strong connection between the intervention of Finland as a member of the international community and the rights and welfare of Afghani women was established to justify the military engagement (von der Lippe and Väyrynen 2011). The operation was argued to be targeted at post-conflict reconstruction in Afghanistan, yet it consisted of a military component.

The concern of the Finnish engagement was founded on the recognition that unlike in traditional peacekeeping missions, in the new theatres of war such as the war in Afghanistan peacekeepers' bodies are blown apart, receive wounds, are stitched together, suffer. In more general, in warfare bodies are disposable objects that are movable, usable and can be shown their proper place and made suffer as Ada's and Bertha's narratives in the previous chapter (Chapter 3) demonstrate (cf. Cornish and Saunders 2014, 1–8; Sylvester 2013, 484). War and violence necessitates bodies in three ways. First, bodies provide the sheer material evidence of the outcome of the conflict and, second, bodies that are touched by it through dead and injury provide the physical evidence of the ideals and ideologies in whose name the war is fought (Stokes 2015, 121). Third, as Elaine Scarry (1985, 60) writes in the context of nation-states, the soldier must be willing to "loan his body to the state for the purpose of war". According to her (Scarry 1985, 112), during wartime the nation "literally inscribes itself on the body" and, thereby, the connection of war and body is intimate as war requires bodies and bodies, in turn, are marked by war.

In addition to the very close and intimate connection between war and body, bodies become touched by war also at a farther distance. Sylvester (2013, 496) writes:

> Others engage with war at a farther distance but in very active and material ways, by producing war material, rehabilitating war-injured people, researching, writing or creating artistic works about war or technological enhancements of war capabilities; state leaders and politicians are also in this category, as are many war protesters. The most physically removed experiences feature about going about normal daily business that seems little changed by war anywhere; these spectators see war clips on television, read about wars in media, play at war through video games and other war-mimicking activities that might be 'innocent' or entail criminal activities, or study wars in school as part of the normal curriculum.

What unites these ways, however, is the human body, a sensing material-physical entity that can touch war and can be touched by war and violence in a variety of ways, also at a distance.

Radically Open Matter and Body

What it means to be touched by war? How we come to be touched by war, conflict and violence? And how conflict bodies can have agency that touches us? Karen Barad's (2003) new materialism and her way of theorizing bodies help to think through these questions. While drawing on the philosophy-physics of Niels Bohr, Barad rejects the atomistic metaphysics of representationalism where the distinction between the represented and the representation is made. She criticizes this binary ontology as it assumes discrete things with inherently determinate boundaries or properties in need of representation. Instead, she expands on Judith Butler's theory of performativity asserting that the world comes into being through iterative everyday performance. Like Elise Boulding's ontology introduced in Chapter 2, Barad's ontological stance shifts the focus from representations to matters of practices/doings/actions. In Barad's (2003, 810) thinking the key question in relation to body to be asked is "how might we understand not only how human bodily contours are constituted through psychic processes but how even the very atoms that make up the biological body come to matter and, more generally, how matter makes itself felt?". In line with the general aim of this book, this chapter founds itself on her onto-epistemological stance where the boundaries of the knower and the known dissolve.

As indicted above, for Bohr things do not have inherently determinate boundaries. It follows that the Cartesian belief in the inherent distinction between subject and object, and knower and known, does not hold in his epistemology and ontology. Bohr's empirical findings in the domain of atomic physics rely on the recognition that any measurement of atoms using an (research) apparatus cannot be attributed to some abstract independently existing object but rather is a property of the phenomenon, namely the inseparability of object observed and agencies of observation. Therefore, according to Bohr, the primary epistemological unit is not independent objects with inherent boundaries and properties but rather *phenomena*. This is the so-called uncertainty principle in quantum physics that is not however, according to Barad, a matter of uncertainty but rather of indeterminacy. In Barad's further elaboration of Bohr's insights, apparatuses are not mere static arrangements in the world, but they are dynamic (re)configurings of the world, specific practices and performances through which specific exclusionary boundaries are enacted. On this agential realist expansion of Bohr's theory, as Barad herself calls her theory, "phenomena do not merely mark the epistemological inseparability of 'observer' and 'observed'; rather, phenomena are the ontological inseparability of agentially intra-acting 'components'" (Barad 2003, 815). For Barad, the term "intra-action" indicates a profound conceptual shift as it is through specific agential intra-actions that the boundaries and properties of the components of phenomena become determinate.

In quantum field theory both human and nonhuman matter is always already a dynamic field of matterings. Matter is not only iteratively reconstituted through its various intra-actions, but radically open. Barad (2012, 7) writes on this mattering and its indeterminacy:

> Ontological indeterminacy, a radical openness, an infinity of possibilities, is at the core of mattering. How strange that indeterminacy, in its infinite openness, is the condition for the possibility of all structures in their dynamically reconfiguring in/stabilities. Matter in its iterative materialization is a dynamic play of in/determinacy. Matter is never a settled matter. It is always already radically open. Closure cannot be secured when the conditions of im/possibilities and lived indeterminacies are integral, not supplementary, to what matter is.

For Barad, bodies are discursive-material phenomena that are characterized by radical openness and they become constituted in intra-action. Bodies include the mutual entailment of the material and the discursive. Barad argues that there is a large number of material-discursive forces, including ones that get labeled, for example, "social", "cultural", "natural", "geopolitical" and "biological", that are important to particular (entangled) processes of materialization of bodies. Bodies are a product of boundary-producing practices as any phenomena. Barad writes (2003, 822) that "discursive practices and material phenomena do not stand in a relationship of externality to one another; rather, the material and the discursive are mutually implicated in the dynamics of intra-activity."

Barad's agential realism also re-conceptualizes agency. Barad emphasizes that agency is not something someone possesses nor is it intentional in the traditional meaning of the word. Agency, according to Barad, is a matter of intra-action and an enactment, hence not something that someone or something has. Barad elaborates agency by saying that agency is not held, it is not a property of persons or things. Rather, it is an enactment, a matter of possibilities for reconfiguring entanglements. Agency is doing and being in its intra-activity. In this theory, agency is not about choice in any liberal humanist sense but "it is about the possibilities and accountability entailed in reconfiguring material-discursive apparatuses of bodily production, including the boundary articulations and exclusions that are marked by those practices" (Barad 2016). Ultimately, for Bard agency is also about possibilities for worldly re-configurings as there is no closure in agency as it is always characterized by radical openness. She (Barad 2003, 827) writes:

> Agency is about the possibilities and account-ability entailed in reconfiguring material-discursive apparatuses of bodily production, including the boundary articulations and exclusions that are marked by those practices in the enactment of a causal structure. Particular possibilities for acting exist at every moment, and these changing possibilities entail a responsibility to intervene in the world's becoming, to contest and rework what matters and what is excluded from mattering.

Barad theorizes also touch in the frame of her agentive realism. Touch for her is coming in contact with something in a way that enables response-ability. Touch is intra-action that is moving and affective. In a similar vein, Sara Ahmed (1997, 27–28) points out that touch has a dualistic resonance in the English language. It can entail the physical contact of your body with another surface or the contact of another skin with yours, but it can also describe the sensation of being affected or moved by something. For Ahmed, being touched suggests becoming closer to each other in which movement across the division of self–other may take place although this movement does not abolish the entire division. In the Baradian thinking, which is in concert with feminist peace epistemology introduced in Chapter 2, I decenter and deconstruct the "me" in the act of touching and being touched as touch brings me close with alterity, both of the self and the other. There is no touch in the singular, argues Erin Manning (2007), as touch is always to share, to open up to something new.

If touch engages the body in moving and affecting intra-action while creating something new, what happens when body is touched by violence and conflict? Body that is touched by violence is, in Manning (2007, 10) words, a wounded "body that opens unto itself and unto others". When this still radically open, yet wounded, body opens unto others in intra-action, the touch of violence circulates between bodies and opens bodies to new directions as violence ties us together since "I cannot affect you violently: I affect us", as Manning (2007, xxx) states. Furthermore, since body is loaned for the purposes of warfare, it becomes a container, a mediator, or even an apparatus of violence (Manning 2007, 66). As indicated earlier, body both receives and disseminates violence.

War Entanglements and Intra-action

The touch of violence at close distant produces "border condition" experiences, experiences of being "in-between" spatial and temporal spaces as the fictional quote in the beginning of the chapter from a female war veteran demonstrates. In a similar vein, a Finnish male soldier, Markus, reminisces his stay in Afghanistan and describes the body that is loaned for the purposes of warfare, that is radically open and experiences alterity.

> Firefight, landmines and grenades were a daily part of being in operation in Afghanistan. One is there to shoot or get shot. The amount of alertness is extremely high, yet it is difficult to always know for what exactly. The feeling of danger is palpable as one is ready to spot, for example, a suicide bomber in a crowd at any moment. When leaving the camp there is 75% risk of getting into a serious danger. Outside the camp chaos prevails. Inside, on the other hand, things run smoothly and in an orderly, military, manner. The soldiers try to make their living quarters homely, although next to one's bed there is a weapon. It is stinky outside since the clay huts are heated with manure. The variation between the daytime and nighttime temperature is extreme. (Pajari 2014)

Heikki, another male soldier, describes the stress:

> The situation is so stressful that many break mentally down. One can tell from the body language, for instance, from the way one walks, that a person is at a verge of mental breakdown. (Pajari 2014)

The life outside the camp is in the stark contrast to the life in barracks that appears orderly, and even homely to an extent. Yet, the body of the soldier is located both inside and outside, i.e. in different spatialities and in the disjuncture of the variety of temporalities. A strong sense of unhomeliness emerges from this border condition. The unhomeliness of the quality of the environment corresponds with the unhomeliness felt in the body. The body senses the pungent odours, hears the noise in the street, feels the changes in the temperature. Aural, visual, kinesthetic and olfactory organs function in the extreme. High sensory awareness prevails as the whole nervous system is in a constant state of urgency and alertness. There is always something to which attend.

Terror touches and enters to the body in the form of unexpectedness of the environment outside the camp. It is felt to be a strange and dangerous terrain where fear and uncertainty looms. The risks are unpredictable. Bhabha's (1992, 148) writing on postcolonial history captures the binary division between the inside and outside when he writes: "These spheres of life are linked through an 'in-between' temporality that takes the measure of dwelling at home, while producing an image of the world history". In terms of temporality, the Bergsonian (Bergson 1910, 100) pure duration (*la durée*) where time "refrains from separating its present from its former states" is felt in the outside as incomprehensible chaos is constitutive of the outside. Yet, it is also a warzone where high technology warfare is waged. When the body enters to the outside it experiences the blurring of the distinction between the past and the present as the present moment brings the past to the present to form an organic whole. In the outside time loses its measurable and linear qualities and the multiple layers of possible pasts actualize in the present of the outside. In short, the duration of the outside is that of the archaic cyclical temporality that is felt to be disjointed from the linear temporality of the inside, namely camp.

A female soldier Elli describes her experience of the intensive sense of pure duration and the spatial disorientation it causes. For her being in Afghanistan presents the "unhomely of history as a disruptive presence" (O'Riley 2007, 1). The colonial past returns creating "a signifying time for the inscription of cultural incommensurability where differences cannot be sublated or totalized" because they occupy the same space (Bhabha 2004, 254). Elli ponders:

> One can loose one's life there. In Afghanistan it is possible that your workmate dies. […] If we can keep the Talibans out, children will be able to go safely to school. […] We tend to forget how much we have helped them and how much there is work left. There would be less hunger, tears and migrants in the world if the wealthy nations were willing to help […] I have never seen such a beautiful night sky with all the stars. The country has rich tradition of culture, literature and poetry which all are lost in wars. (Karhula 2012)

The corporeal choreographies that include the Afghans resemble the encounters of the colonial type producing affective and interruptive moments that become haunting due to their incommensurability.

The country feels as remaining "haunted by the memories of imperialism, doomed to linger in a time lag that condemns it to repeat the conflicts of its colonial past [...]" (O'Riley 2007, 4). Tuomas narrates the choreographies in the following manner:

> The soldiers spent a lot of time in the villages talking to the locals. The mixture of odors was extraordinary: garbage fires produced black smoke, open sewages smelled urine and the sweet smell of horse and donkey stool floated in the air. The traffic was chaotic as horse carriages, donkeys, bicycles, pedestrians and cars congested the streets. Turban-headed old men were sitting and drinking tee in the street corner cafes. They watched the passing by patrols with a great suspicion in their eyes. (Muraja 2014)

Appearance of the local men is incorporated in an affective register of danger and terror. The metamorphosis from a turban-headed body into a "Muslim" via the mediation of "terrorist" through the chaotic intensity of the sensory experiences takes easily place. In this register, "terrorists" appear extremely effective because they infiltrate the locals with all the performative accoutrements of "normal" behaviour of, for example, drinking tee in the street corner cafes. Terror is high since the "terrorist" is on the constant move and process of transformation and infiltration. Outdoing this turban-headed body is not possible as signs of danger are opaque and lend themselves to be easily misread (Bharucha 2014, 74–110). In this register, turban-headed body has intensive (colonial) affectivity in the state of high sensory awareness as fear "materializes in the turban" and "loaded tactile economies" transform the turbaned-body into the terrorizing body (Puar 2007, 94, 187).

Heikki narrates how terror makes line between the agent and object waver (cf. Spivak 2004, 92):

> As the locals resisted the military mission, the danger grew larger: roadside mines, armed attacks and suicide bombs became a common occurrence. A stone in a wall could have been a target of a rocket attack. Soft drink can, helmet or a plastic bag could turn out to be a bomb. Someone talking nervously to a mobile phone could mean that the phone was a remote control. If the streets were empty and people were ushering children to move away, something was certainly going on. (Pajari 2014)

The wavering of the line results in fear and uncertainty in the "outside". The dread of ordinary and homely, e.g. a stone, soft drink can, plastic bag and a mobile phone, produces the collapse of that what was considered to be rational. A vivid displacement follows as things appeared once *heimisch*, homely and familiar, turn out to be the opposite (cf. Bhabha 1992, 148–150). A shift of corporeal focus takes place as these banalities transform the environment a location of terror and danger. Ordinary material things represent now the outsideness in a terrorizing manner. They freeze the world into an unhomely location, into another world, which creates an arrhythmic resonance in one's body.

Bringing War Home

The Baradian frame of new materialism enables to understand how the war body circulates violence across bodies, spaces and times. Allen Feldman (1991, 7) concurs both with Barad and Manning when he states that the body "altered by violence, re-enacts other altered bodies dispersed in time and space; it also re-enacts political discourse and even the movement of history itself". What is typical of violence is that it "circulates from one surface to another, from one space to another, from one body to another, from one space-time to another" (Manning 2007, 67). In the framework of body that is touched by war, this circulation means that corporeal matter touched by violence, makes itself felt in another body in a manner that expands across space and time as well as across surfaces. Corporeal matter has war agency, and hence agency to bring distant warfare home when it becomes felt in other bodies dispersed in time and space. It is the indeterminacy of mattering described earlier in this chapter that allows the expanding and boundary-crossing circulation of the touch of war to happen.

Bodies circulate the touch of war in their unique ways. Markkus' body withdraws from the family activities because he becomes irritated easily.

> When I came home for holidays, I was absent minded and got easily irritated. Children's running around home disturbed me a lot. In Finland, I am in a grey zone. In Afghanistan, things are black and white. (Pajari 2014)

The body that is at home emerges as an unhomely presence. The homely "banalities"—the sounds of playing children, getting sleep, paying attention to mundane duties—are fundamentally stirred. Strangeness falls on the details of life: how you can sleep, how much you can take from your family, whom you can talk to are not certain any more. The war has forcibly entered the everydayness of life to dispossess, divide, alarm, interrupt, namely, to haunt. The body is located in this familial world, but not fully a part of it. The outsideness has entered to the inside.

The body in the mission has bore witness to many things that later remain with it. Witnessing refers to the Latin word *superstes* which "designates a person who has lived through something, who has experienced an event from beginning to end and can therefore bear witness to it" as Christopher Harker (2007, 53) writes in reference to Giorgio Agamben. Witnessing itself is always a bodily act that requires corporeal presence. The war veterans bear witness something that exists in different time-space, and it is their movement between different locations and across temporalities that eventually allows them to bear witness as well as their witnessing to be felt and circulated in other bodies.

Tuomas describes how having lived through the war and violence disturbs his subjectivity, decentres him and forces him close to the alterity in himself:

> After all the bad things you have witnessed, maybe it is yourself that you come afraid of. (Pajari 2014)

The intersection of time-space allows new entanglements of war to emerge: it is this "interstitial geography", "an in-between spacing that brings 'here' and 'there', [...], into a common space through [her] act of witnessing" as Harker (2007, 58) describes witnessing violence. He continues, in concert with feminist peace theorizing and its epistemology introduced in Chapter 2, that in witnessing it is impossible to establish the position of a stable subject as the "witness-subject is always already in motion, an intersubjective process taking place between things and people" (Harker 2007, 66).

Summary

Space of post-conflict peacebuilding is filled with bodies that are touched by war and violence in a variety of ways war as violence touches some bodies directly and others across spatial and temporal boundaries when material bodies circulate violence across time and space. Through the case study material I have discussed in this chapter how the touch of war and violence is brought home in the bodies of Finnish war veterans.

The Second World War engaged Finland in territorial warfare close to its borders, whereas the war in Afghanistan was a distant war where the Finns took part, first time ever, in an international military mission. Both of them indeed brought war and violence home, but they brought it differently. In the Second World War, the number of wounded and dead Finns was large and war touched closely most of the Finnish families as indicated in Chapter 3. The war veteran's bodies had a strong affective presence among the Finnish population and the war was felt in multiple ways for decades. Finland has not take part in the colonizing wars, yet the war in Afghanistan and war veterans' bodies has brought post-colonial history and present condition to Finland as the bodies touched by war in Afghanistan has became felt in and through multiple agential intra-actions.

I have argued in concert with the Barad (2003) new materialism, that the radical openness of matter brings forth agency that is about the possibilities of reconfiguring of boundary articulations. In the concluding chapter (Chapter 8) I will discuss the implications of this openness for peacebulding as an response-able intervention to violence. In this chapter, on the other hand, I have sought to demonstrate that the boundary between us and them becomes reconfigured as the witnessing body circulates the touch of war across time and space due to its radical openness. This chapter has underscored that different entanglements of time-space and subjectivities emerge through witnessing when "us here" encounter "them there" in an intersubjective and corporeal manner (cf. Harker 2007). Harker's reading of witnessing enables a Baradian interpretation as he writes that witnessing is always intersubjective and this "intersubjective relation occurs not only between 'human' subjects, but also with a whole variety of other 'things': videotapes, detention centres [...]" (Harker 2007, 69). The witnessing narratives of, for example, the war veterans establish corporeal and intersubjective relations where the touch of violence and war is mediated and transferred in and through the body of the veteran.

References

Ahmed, S. (1997). Intimate Touches: Proximity and Distance in International Feminist Dialogues. *Oxford Literary Review, 19*(1–2), 19–46.

Barad, K. (2003). Matter Feels, Converses, Suffers, Desires, Yearns and Remembers. *Signs: Journal of Women in Culture and Society, 28*(3), 801–831.

Barad, K. (2012). On Touching—The Inhuman That Therefore I Am. *Differences: A Journal of Feminist Cultural Studies, 23*(3), 206–223.

Barad, K. (2016). Matter Feels, Converses, Suffers, Desires, Yearns and Remembers'. Interview with Karen Barad, interview by Rick Dolphijn and Iris van der Tuin. *New Materialism: Interviews & Cartographies.* Accessed May 30, 2018. http://quod.lib.umich.edu/o/ohp/11515701.0001.001/1:4.3/–new-materialism-interviews-cartographies?rgn=div2;view=fulltext.

Bergson, H. (1910). *Time and Free Will: An Essay on the Immediate Data of Consciousness* (F. L. Pogson, Trans.). London: George Allen and Unwin.

Bhabha, H. (1992). The World and the Home. *Social Text, 31*(32), 141–153.

Bhabha, H. (2004). *The Location of Culture.* New York and London: Routledge.

Bharucha, R. (2014). *Terror and Performance.* New Delhi: Tulika Books.

Butler, J. (2004). *Undoing Gender.* New York and London: Routledge.

Cornish, P., & Saunders, N. (2014). Introduction. In P. Cornish & N. Saunders (Eds.), *Bodies in Conflict: Corporeality, Materiality, and Transformation* (pp. 1–8). London and New York: Routledge.

Edkins, J. (2003). *Trauma and the Memory of Politics.* Cambridge: Cambridge University Press.

Enloe, C. (1993). *The Morning After: Sexual Politics at the End of the Cold War.* Berkeley: University of California Press.

Feldman, A. (1991). *Formations of Violence: The Narrative of the Body and Political Terror in Northern Ireland.* Minneapolis: University of Minnesota Press.

French, L. (1994). The Political Economy of Injury and Compassion: Amputees on Thai-Cambodia Border. In T. Csordas (Ed.), *Embodiment and Experience* (pp. 69–99). Cambridge: Cambridge University Press.

Harker, C. 2007. "A Close and Unbreachable Distance": Witnessing Everything and Nothing. *ACME: An Internationsl E-Journal for Critical Geographies, 6*(1), 51–72.

Karhula, R. (2012). Karu Afganistan tarjosi rauhanturvaajalle tähtitaivaan. YLE 4.12.2012. Accessed May 29, 2018. https://yle.fi/uutiset/3-6400496.

Kivimäki, V. (2012). Between Defeat and Victory: Finnish Memory Culture of the Second World War. *Scandinavian Journal of History, 37*(4), 482–504.

Lunde, H. (2011). *Finland's War of Choice: The Troubled Finnish-German Alliance in World War II.* Havertown, Newbury: Casemate Publishers.

Manning, E. (2007). *Politics of Touch.* Minneapolis: University of Minnesota Press.

Muraja, T. (2014). Afganistan ei jättänyt Juhania rauhaan. *Apu.* Retrieved May 29, 2018, from https://www.apu.fi/artikkelit/afganistan-ei-jattanyt-juhania-rauhaan-han-paatti-ampua-itsensa-miten-suomi-huolehtii.

O'Riley, M. (2007). Postcolonial Haunting: Anxiety, Affect, and the Situated Encounter. *Postcolonial Text, 3*(4), 1–15.

Pajari, K. (2014). Afganistan jätti jäljen – rauhanturvaajalla on veteraanin kokemus, mutta ei titteliä. *Helsingin Sanomat, 29*(12), 2014.

Pentley, J. (2003). War Without Death: Responses to Distant Suffering. *Journal for Crime, Conflict and Media, 1*(1), 72–85.

Puar, J. (2007). *Terrorist Assemblages: Homonationalism in Queer Times.* Durham and London: Duke University Press.

Sanghera, G., Henry, M., & Higate, P. (2008). *Peacekeepers as New Men? Security and Masculinity in the United Nations Mission in Liberia* (Working Paper No. 02–08). School of Sociology, Politics, and International Studies, University of Bristol.

Scarry, E. (1985). *The Body in Pain: The Making and Unmaking of the World.* New York and Oxford: Oxford University Press.

Spivak, G. (2004). Terror: A Speech After 9/11. *boundary 2, 31*(2), 81–111.

Stokes, J. (2015). Fragments of a Great Confusion: Abjection, Subjectivity, and the Body in Mary Borden's *The Forbidden Zone.* In K. Sellberg, L. Wånggren and C. Aghtan (Eds.), *Corporeality and Culture: Bodies in Movement* (pp. 115–126). New York: Routledge.

Sturken, M. (1991). The Wall, the Screen, and the Image: The Vietnam Veterans Memorial. *Representations, 35,* 118–142.

Sylvester, C. (2011). Experiencing War: An Introduction. In C. Sylvester (Ed.), *Experiencing War* (pp. 1–5). London and New York: Routledge.

Sylvester, C. (2012). War Experiences/War Practices/War Theory. *Millennium: Journal of International Studies, 40*(3), 480–503.

Sylvester, C. (2013). *War as Experience: Contributions from International Relations and Feminist Analysis.* New York: Routledge.

Vannini, P. (2015). Non-representational Ethnography: New Ways of Animating Lifeworlds. *Cultural Geographies, 22*(2), 317–327.

Väyrynen, T. (2016). The Finnish National Identity and the Sacrificial Male Body: War, Postmemory and Resistance. *National Identities, 18*(2), 217–232.

Väyrynen, T. (2014). Muted National Memory: When the "Hitler's Brides" Speak the Truth. *International Feminist Journal of Politics, 16*(2), 218–235.

von der Lippe, B., & Väyrynen, T. (2011). Co-opting Feminist Voices for the War on Terror: Laura Bush Meets Nordic Feminism. *European Journal of Women's Studies, 18*(1), 19–33.

CHAPTER 5

Abjected and Silenced Bodies

Abstract Some post-conflict bodies carry historical and political content that cannot be reconciled with peacebuilding. Peacebuilding is a site of struggle where some bodies and narratives are casted out from the core of the community as the bodies that remind the community of its suppressed past are rejected and silenced. This chapter demonstrates how the bodies of women who fraternized with the German soldiers during the Second World War are post-conflict bodies that became cast out from the core of the nation during peacebuilding. The chapter highlights that despite the rejection, post-conflict abject bodies can become *parrhesiastes*, speak back to power, and retain the right to remind the post-conflict collective self of the residual historical content that bears upon the present.

Keyword *Parrhesiastes*

"Old German slut, that was the word they used to call me. Not in front of me, of course, but behind my back. I know that here in a small village they call me that", laments an old woman, Elma, in a Finnish documentary entitled *Auf Wiedersehen Finnland* (2010) by the Finnish documentary film maker Virpi Suutari. Elma's voice is high-pitched, but strong. She is not making any excuses, because for her there no longer

T. Väyrynen, *Corporeal Peacebuilding*, Rethinking Peace and Conflict Studies, https://doi.org/10.1007/978-3-319-97259-6_5

anything to apologize for. Her long silence has been brought to an end and nothing is going to stop her from speaking now.

Post-conflict society is a site of struggle where some bodies and narratives are casted out from the core of the community. In short, peacebuilding is by no means an unambiguous, harmonious and non-political process that follows a linear path. Rather, it creates differences and closures. For example, Elma's quote above demonstrates that her body is in an intimate relationship to the post-conflict community that, however, has deserted it. Her presence, on the other hand, presents an exposure that reveals hers—as well as ours—openness and vulnerability to the world and to the others, and thereby demonstrates the fragility of peacebuilding. Her body has been lived through by a variety of forces (cf. Barad 2012, 218) during the war that cause her dismissal and, yet, her body that has experienced violence touches us and forces us, even after decades, live through forces that constitute her abject existence. This chapter examines narratives and corporeal performances of persons like Elma who have witnessed war, but who have been abjected during peacebuilding. By doing this, I shift the attention in this chapter to the practices of governance that control bodies during war and post-conflict peacebuilding.

In order to understand the mechanisms of governance that are targeted at bodies during war and peacebuilding, I study in this chapter such instruments as naming, racialization, medicalization and securitization targeted particularly at female bodies. These mechanisms are studied in the context of the Second World War Finland, and yet these mechanisms function in many post-conflict societies who seek to reconcile with the past war and violence during the process of peacebuilding. The case study material I use relates to the Finnish women who fraternized with alien soldiers, namely the German army during the Second World War. In order to study the mechanism of silencing and control I use material that has only very recently come to light, i.e. the first historical studies and a documentary film that explore the narratives of the women. I attempt a feminist reading of the material and seek to problematize their disciplined bodies in the wider context of post-conflict peacebuilding. Like in the third chapter (Chapter 3) of this book where the relational and sensuous bodies' capacity to temporarily escape the strict regime of war and its disciplinary control, resistance characterizes also the bodies I examined in this chapter.

As noted above, I employ feminist theorizing to show that female bodies become encoded as the symbols of the nation and its purity, and that sacrificial male bodies are, on the other hand, symbolically coupled with the nation's heroism and destiny (Butalia 1998; Das 2000, 2008; Ehlstain 1987; Enloe 2000; Hutchings 2008; Jarvis 2004; Mosse 1996; Nagel 1998; Yuval-Davis 1997). The European experience of the German occupation during the Second World War, and particularly female collaborators' stigmatization as well as "comfort women" in Asia bear similarities with the case study material presented in this chapter (Aarnes 2009; Ericsson 2011a; Ericsson and Simonsen 2005; Gildea et al. 2006; Hayashi 2008; Soh 2008; Virgili 2002). Feminist theorizing on war is useful in this contexts since it has drawn attention to women's corporeality in symbolizing the nation and its honour particularly in times of violence and war (Butalia 1998; Ehlstain 1987; Enloe 1990; Yuval-Davis and Anthias 1989). The abject figures of, for example, "comfort women" who have been subjected to nationalistic violence and "female terrorists and torturers" taking part themselves in violence and war have been also brought to light through empirical cases studies in the feminist Peace and Conflict Studies (Parashar 2011; Sjoberg and Gentry 2007).

This chapter, however, departs company from most of the above-mentioned literature by using explicitly psychoanalytically oriented reading of the research material. In short, it examines the post-conflict nexus of gendered bodies and peacebuilding nation and, thereby, seeks to show how defining the collective identity during peacebuilding and the production of its imagined coherence require female bodies that are appropriated, abjected and even muted. I am particularly interested in "not so honourable female bodies" and seek to show how the disciplinary mechanisms of abjecting and silencing work in peacebuilding contexts.

This chapter is also about the voices silenced post-conflict bodies produce. I rely on Mladen Dollar's (2006, 73) work, where she argues that "what language and the body have in common is the voice, but the voice is part neither of language nor of the body". In short, voice cannot solely be reduced to language or body. Bodies can, however, be muted by categorizing bodies, ejecting some bodies to the nation's margins (e.g. lunatics, the disabled, prisoners, dishonoured women) and excluding their "small voices" (Guha 1999, 3) from the hegemonic history writing of the nation. In order to avoid producing the category of victim, the chapter shows how manifold memory formations and voices enter to peacebuilding space after violent conflict and how they contest the hegemonic forms of post-conflict national narrative, and eventually give rise to post-conflict agency. The agentative figure that emerges is not that of a *superstite* (survivor) witness with confessional tendencies as the war veterans are in Chapter 4 are, but that of a *parrhesiastes*, the one who speaks back to power.

My diffractive reading of the material is also guided by Slavoj Žižek's (2006, i–ii) idea of a "short circuit" which occurs when we cross wires that do not usually touch. For Žižek, this is one of the most effective critical procedures because it shatters and undermines our common perceptions. In the Žižekian spirit, the chapter seeks to cross lived corporeal experiences, feminist Peace and Conflict Studies and psychoanalytically oriented political theory with performative and acoustic case study material. Since the focus is on the female body as well as its performative and acoustic dimensions, I use the unique visual material available in Suutari's documentary where four old women tell their stories of their fraternization with Nazi soldiers, travelling to Germany and back to Finland. The overall argument I present is psychoanalytical and figural, not as such context-specific, and the documentary is read without discussion of the film's status as evidence. In short, I examine the corporeal representations and voices of the women not as simple representations of a past event, but as post-conflict social and political performances and utterances which intervene in a past and present peacebuilding context.

From the perspective of corporeal peacebuilding, it is important to note that war and violent political conflicts are some of the most totalizing events in which a large-scale categorization of human bodies takes place as also indicated in Chapter 3. In war, the "scripts of nationalism" (Das 1998, 68) are violently written on human flesh and violence touches bodies in multiple ways. Furthermore, a variety of extreme disciplinary mechanisms, e.g. surveillance, naming, medicalization and examination, are targeted at human bodies during conflicts and the disciplinary control intensifies during intensive political violence. In psychoanalytically inspired reading of violent conflict, conflict and war are traumatic for a whole nation or a community as they reveal the ways in which the collective identity is always radically incomplete and fragile. War and violence disrupts the delicate collective trust in the practices of meaning that constantly produce the imaginary of coherent and stable collective identity (Shapiro 1997; Edkins 2003, 2006; Fierke 2004; Bleiker and Huchinson 2008; Väyrynen 2013). It shatters the existing structures of identity by forcing the nation or community to confront "an event that, in its unexpectedness and horror, cannot be placed within the schemes of prior knowledge" (Caruth 1995, 153).

The trauma of violent conflict and war leads to a heightened sense of collective anxiety. An urgent need for post-violent closure emerges during peacebuilding and "enormous ideological energy then will be used to stitch this tear in the symbolic fabric back together" (Berger 1999, 25) in order to create the collective fantasy of an unitary community. This anxiety is extremely intensive during peacebuilding as there is a struggle over the collective identity. It follows that post-war collective anxiety creates sealed collective memoryscape where residual historical content is shunned away. The bodies that remind the community of its suppressed past are rejected from the core of the community and differences and social boundaries are created. In this chapter I demonstrate how the figure of *parrhesiastes* (the one who speaks the truth)—not just the figure of *superstes* (survivor witness) discussed in Chapter 4—is also needed to transfer the community from conflict to peace.

Controlling Female Bodies

During violent political conflicts home front is produced as a moral and political site whose defence is almost as important as the defences of the military front. As outlined in Chapter 3, the guarding of the moral superiority of the homefront requires disciplinary measures which quarantee that bodies follow the rules set during the war. As Ada indicated in that chapter, the movement of bodies were monitored in multiple ways: dances were forbidden, nightime black out restricted the bodies' movement in towns and villages and the encounters with the alien soldiers were assumed to follow a pre-written choreography. Particularly female bodies were targeted by surveillance and control of the moral behaviour. During conflict any rupture in the unity of home front is seen to imperil the entire community, and thereby surveillance, control and identifying and naming of the outcast subjectivity becomes a means of discipline. As the imaginary work that produces the assumed coherence of the collectivity largely takes place at the home front the uncontrollable dangerous energy of, for example, the sexual desire of the female body needs to be harnessed right there.

In Finland, the home front became under extreme surveillance during the Second World War when 200,000 alien military men traversed or were stationed in Finland. The civilian—alien soldier choreographies are often tightly controlled and monitored during conflicts. The Finnish-German encounters and interaction that took place in the everyday life in the towns and villages were initially neutral or even friendly as Ada's war story in Chapter 3 demonstrates. The general attitude of the Finnish population was by no means hostile as the Germans represented the country's allied force against the Soviet threat. Many of the Germans soldiers stayed for a long time in Finland and intimate relationships with Finnish women developed (Virolainen 1999, 77–94; Junila 2000, 138–143, 286–304).

In most of the conflicts active measures are taken to control the behavior of the women in home front. As indicated, female sexuality becomes a source of major anxiety and the intensive practices of, for example, naming through which the "confluence of historical, cultural, biographical, political, and symbolic themes that express membership in a particular group" (Tanno 1994, 33) brew up. In Finland, the wartime Information Bureau was active in reporting on relations between Finnish civilians and German soldiers (Westerlund 2011, 325–344). The reports were in the beginning cautious about emphasizing in a negative tone the relations between Finnish women and German soldiers, but the tone changed as the concern about the loose morality of the Finnish women emerged. "The rumor that the Finnish women are overtly excited about the Germans spreads among the population", reports the bureau in late-1941 (Westerlund 2011, 339). Naming contributed to the discursive creation of (im)moral subjects in the home front:

> [...] the decline of the morality during the wartime rightly begets serious concerns as it looks as if the population thinks that it is free from moral constraints. People do things they would not otherwise do. Drinking and promiscuity have increased. The reason for the moral decline can be found in the deployment of the German troops, but there are other reasons, too. Let us pay attention to the fact that many war widows have turned to prostitution. There are cases were women have abandoned their children and now run loose in the town and elsewhere. (Westerlund 2011, 339–340)

The quote above indicates how the idea of "womenandchildren" (Enloe 1990) and their location at the nation's core form a core signifier through which female bodies can be named and disciplined.

The presence of the "contaminated" bodies of foreign soldiers constitute a danger to the "healthy and vigorous body" of the nation in conflict. To avoid the "contamination" of female bodies extreme forms of disciplinary measures in the form of, for example, medicalization can be initiated. This happened also in Finland when the loosening of morality was seen to be accompanied with a threat of venereal diseases (Westerlund 2011, 336–340). In general, medicalization follows a structure where, first, there is an identification of the pathogenic situation. Second, an inventory of the deviant situation follows and finally, a set of prescriptions administered to remedy the situation. It involves expert knowledge which has the right to diagnose, name and prescribe the cure (Lock and Kaufert 1998; Riessman 2003; Feldman 2004, 170). As Maya Todechini (2002, 124) notes in the context of Japanese *hibakusha* women that the "transformation of *hibakusha*'s bodies into a series of medical and legal data, neatly circumscribed by experts, can also be seen as a subliminal collective attempt to control bodies that in fact were far more threatening than was publicly acknowledged".

Women carrying venereal diseases often became a ready-made target for the medical gaze. In this gaze moral and medical pathologies become conflated and the control of venereal diseases become a controlling mechanism for female bodies (Davidson and Hall 2001). Women who transmit the diseases are seen to be loose and represent a threat not only to the nation's health but, most importantly, to its morality. In Finland during the Second World War, venereal disease was a signifier in which the nation's honor culminated. The German medical officials reported to the Finnish authorities the amount of venereal diseases transmitted from Finnish women to German soldiers (Sauerteig 2002; Timm 2005). Detailed documentation of women who were suspected of transmitting diseases was then collected by the Finnish medical authorities including the women's names, occupations, addresses and even behavioural characteristics. The information was filed in the archives of the Finnish Medical Bureau and employed in several occasions as a disciplinary mechanism (Junila 2000, 254–258).

Doubly Outcast

The racial context and racialization, which is a common feature of many of the current conflicts and wars, characterized also Finland during the Second World War. In Finland, many Finnish-German couples wanted to marry, but marriages between German soldiers and Finnish women required complicated bureacratic procedures. Eventually, a few managed to obtain marriage licences. Eduard Dietl, the German commander of *Alpenjäger* in Lapland, wrote about the marriage applications:

> With very a few exceptions, the applications that have been submitted unfortunately involve representatives of neighboring peoples of significantly lower value. The pictures shown almost exclusively depict racial driftwood, starting with girls showing strongly eastern features, all the way to an ugly bride of inferior growth. (Markkanen 2011)

Dietl's comment can be read to represent an example of how female bodies were subjected to reproductive control. When the note was written the strict racial categories that dictated with whom Germans could produce children were in force (Ericsson 2011b; Junila 2000; Kundrus 2005; Heiskanen 2011; Westerlund 2011). For Dietl, and presumably for other officers who were granting the licences, the women's external appearances were enough to deem many Finnish women racially, mentally and physically inferior despite the military alliance: the Finnish women had come to signify a corrupting influence on the German race.

Given that marriage licences were difficult to obtain, children were born out of wedlock. Ada pays attention to these children in her story in Chapter 3 with curiosity. The official estimate of the children born out of wedlock vary from 500 to 2000 although there were attempts during the war to list the single mothers in order to guarantee their social benefits from Germany on the basis of the contract between Finland and Germany. Municipal child protection did their best to keep records of illegitimate births which allowed close scrutiny of female bodies in the name of children's well-being. The child protection agencies tried to identify the children's fathers who would then be required to pay for the child's support. Mothers were entitled to name the father. Many women who had relations with German soldiers wanted to avoid being stigmatized and reported the child's Finnish father had been killed in action. There were also women who "were not willing to confess, not even under coercion" as it was put in their files (Junila 2000, 247). Confession within a relationship of power gives the authority demanding the confession a tool by which the individual can be assessed, named and categorized and even corrected (Foucault 1978). If the confession was not provided voluntarily, extreme measures, including psychological pressure, were taken by the Finnish child protection authorities to extract the information. In other words, an obligation to tell the truth about oneself, to confess one's behavioral tendencies, was created in the name of child protection.

The previous chapters (Chapter 3 and 4) have demonstrated how missing bodies and mentally shattered bodies become easily rendered abject. There is a third category of post-conflict abject bodies, namely bodies who had fraternised with the alien soldiers. In Finland, women who fraternized with the German soldier and particularly those who left for Germany with the withdrawing troops became a target of othering. Although their reasons for leaving included love, work, the fear of Russian occupation, social pressure and also adventure, they became labelled as traitors (Heiskanen 2011). They were seen to have betrayed their nation in two ways. First, they were suspected of having fraternized with the Germans and thereby rejected the sacrificial Finnish men and soldiers, and second, they had deserted their nation at a critical point in the war. Their restless bodies, that were ready to follow the aliens, could not be employed to construct the national myth, and thereby a closure to the post-conflict identity narrative of honorable and laborious women, who took part in the Finnish war effort on the home front. They had deserted their country in a traumatic moment and deserved to remain unrecognized and muted by the nation.

Gendered bodies are often also securitized during conflict and peacebuilding. For example, women who had left Finland were interned in a camp on their return to Finland where intensive securitization of their bodies took place, namely their bodies were defined as a security risk in whose name internment and strict interrogation were justified. The Finnish Security Police interviewed 658 women who returned Finland between 1945 and 1948. Many of the women had experienced life in a variety of camps (camps administered by the UN, the Red Cross or the British troops) in which they had survived to find themselves again in a camp. A camp for persons who return to Finland was established by the Security Police in a small town, Hanko, for prisoners of war and returnee civilians from the Soviet Union and later from Germany. The Security Police was particularly interested in any activities of the women possibly amounting to treason (Heiskanen 2006, 2011; Kontinen 2011).

When interrogating for alleged treason the Security Police named and securitized female bodies. It reported in August 1945:

> A record was set by a woman who returned from Germany after having first been engaged to a Finnish man and then fell in love with a German. She followed him to Germany in 1944. There she fell in love with a French man and when he returned to France she found a Pole. When she returned to Finland she thought that she was pregnant either by the Frenchman or the Pole. Yet she carried the signs of English men, too. During her interrogation she admitted most recently courting English men. (Kontinen 2011, 159)

As indicated in the quote, this kind of restless female body who "carried the signs" of alien men became for the post-war Finnish nation a center of intense significations. It became a center of significations in a particular way: the existence of women, who fraternized with the Germans became "unalterable, precise, but enveloped in the skin of memory" (Delbo 1990, 4) of the nation. The nation wanted to cast off its own past in relation to the women, as they possessed knowledge that was considered to be harmful. Through their corporeal existence the past entered the peacebuilding moment as "poisonous knowledge" (Das 2007, 76), with which the nation would have nothing to do. The nation wanted to discard the women, to create new skin made up of present memory that would cover up the deep skin. However, the skin of memory did not renew itself, although post-war Finnish memoryscape was formed where talking about this part of the national history was socially sanctioned. Ultimately, the living bodies of the women constituted a reminder that hindered the production of closure to the trauma of war, particularly of the Finland's shameful alliance with the Nazis.

Speaking Between Two Deaths

Post-conflict silences can take a long time to crack. Peacebuilding and reconciliation as such do not necessarily generate their cracking. In Finland, the state-initiated disciplinary practices, that led to the banishment of the abject from the core of the collective, contributed to the creation a variety of silences during peacebuilding. These silences started to crack just a few years ago, six decades after the war, when some historians and artists brought a part of this residual historical content back to the nation's consciousness. The material included, for example, war photographs that showed the executions of the enemy soldiers at the war front, research on the maltreatment of the Soviet prisoners of war as well as artwork that dismantled some of the most sacrificial war figures. The Finnish National Archives conducted research on the children of foreign soldiers in 2010 which indicated that also the state have some interest in the material. Some of the silences have been so profound that the new generation of Finnish historians, who wanted to interview, for example, the women who had fraternized with the German soldiers and their relatives did not succeed.

Post-conflict corporeal silences can be internal and external. For some it is an internal silence where the person attempts to suppress what is recalled. For them, the weight of the disciplinary measures and their memories are so overwhelming that suppressing them is a way to maintain a coherent self-image. In some cases the trauma is so severe that it does not break into language (cf. Scarry 1985, 4) . For others, the silence is external: they do not want to tell what they recall because others do not care to hear what is said. The abject is able to speak, but the post-conflict community does not want to hear what the abject wants to say (cf. Butler and Spivak 2007). For some, there is a need to tell, but the intense pressure not to tell silences them. Some people have also a conviction that they would not be believed if they told about their experiences, hence, the silence (cf. Leys 2000; Radstone 2007).

Despite the controlling mechanisms and silencing, bodies can escape the mechanism in multiple ways. Corporeal peacebuilding indicates also space where the mechanisms loose their control and the body finds a new way of being in the world. For example, Elma who was quoted at the beginning of this chapter, has found her voice after having experienced the social death of being categorized as Hitler's bride and being muted by the Finnish nation for sixty years. Veena Das (2007, 62) notes that the theme of a woman possessing poisonous knowledge, but who finds her voice when she is occupying the zone between two deaths, is an important one. In a similar vein, Elma in the documentary can be considered to have found her voice after having first been appropriated, categorized, disciplined and finally abjected by the nation that caused her social death. Second, she can now be seen to be approaching her biological death, hence her intensive and tumultuous existence between two deaths. Das (2007, 62; cf. Guha 1999, 12) continues, that the "truth articulated from this zone is rarely enveloped in beauty or splendor". Elma's truth is not beautiful: she shakes her fist, swears and hisses angrily. "Here you have a woman who has lived", she shouts. Her body is not the hunched or broken body of a traumatized individual, but it is a body that has finally found its voice, and is willing to talk back to the post-war nation and is not ashamed to let the body's truth come out. Her tone is harsh because she can see her life as already lived.

Elma's body is a typical "war body" that is transformed by disciplinary political processes and where bodily memory, biography and national history merge: the past is sedimented in it. As noted earlier in this book, the body is always already embedded in history. In Elma's body, the state-initiated disciplinary processes of naming, medicalization, securitization and non-recognition cross over from the public space to traumatize the inner space and then cross back in the form of collective anxiety. Arthur and Joan Kleinman (1994, 716–717; cf. Scarry 1985, 108–133; Stoller 1994, 638–639; Motsemme 2004, 916) write about this type of transfer and transformation:

> That is to say, bodies transformed by political processes not only *represent* those processes, they *experience* them as the lived memory of transformed worlds. The experience of memory processes is sedimented in gait, posture, movement, and all the other corporeal components which together realize cultural code and social dynamics in everyday practices. The memorialized experience merges subjectivity and social world.

Elma's corporeal components and voice speak her pain as they are repositories of cultural memories. Socially constructed external memory could only skate along the nationally approved collective memories, but not touch the body's re-experience. Her past is not known just in words, but also in her body. It is the "deep memory" that is felt through her whole body. Her body knows the truth that she is finally putting into words. Elma's experiences are communicated and tense signifying practices incorporated in her body are released.

Traumatic experiences do not always leap into words as indicated earlier. In the documentary Roosa's story is an example of not being able to talk, and yet she allows her son, Frans, to narrate their experiences. Frans is a middle-aged man in search of his German father. He was born in Germany in 1945 and moved to Finland with his mother. In his early childhood he was placed in his aunt's care. Perhaps Roosa wanted to avoid the disciplinary practices imposed in the name of child protection. Frans wants to know more about his very early childhood by searching for his relatives in Germany. His mother is no help at all. Frans complains in the documentary that his mother never wanted to talk about the war nor his father. The closing scene of the documentary shows Frans and Roosa sitting next to each other, looking at an old black and white photo, where a beautiful young woman leans slightly towards a handsome German soldier. Roosa had declared a minute ago that there is nothing to talk about, but Frans persists in teasing something out of her. By showing the photo and by evoking a casual conversation with his mother, Frans allows Roosa to circulate the post-war conditions that have rendered her silent for sixty years. Roosa's voice rises and lowers and her face shows an abundance of powerful affects when she declares that "think, we have endured that all".

Silence is not necessarily a marker of passivity, it may imply post-conflict and peacebuilding agency. Roosa's body carries significations that connect her to the nation and its trauma. The price may have been her internal silence, but she has carried it with dignity. Moreover, as Bibi Bakare-Yusuf (1997, 176) writes, "[i]t should, however, be noted that the formative role of language in subject formation is not the only means of constituting subjectivity". Nthabiseng Motesemme (2004, 917; cf. Clair 1997; Dollar 2006; Ross 2001) notes, on the other hand, in reference to Foucault that while silence is important in keeping the grids of power intact, silence also has the capacity to loosen its hold and hence make it possible to thwart hegemonic power itself. Roosa may have wanted to protect her family members by withholding information as an act of conscious resistance. Furthermore, by letting Frans tell their story and by allowing her body to enter into the nation's consciousness in the form of the documentary, she talks back to power.

Post-conflict Memoryscape

Having experiencing a hostile peacebuilding context does not necessarily imply losing one's agency. Terttu's story is an example of a "memoryscape", where the post-conflict nation positioned her body. Her flight from Finland with her German boyfriend took her through the unfriendly northern landscape. During their flight they stopped first in a small Norwegian village, Alta, where the grey and vast sea meets the rough cliffs of the northern shore, as Terttu describes the landscape in a matter of fact voice. The trip continued with a boat through the dangerous and stormy Arctic Ocean. Her journey across the icy sea has a metaphorical quality that has some intangibility as well as concreteness in it. The metaphysical memoryscape, which Terttu creates in her narrative, also describes the peacebuilding nation's response to her actions. The response was harsh, rough and unfriendly. Her internal memoryscape forms a collage where Terttu's memories of the flight meet the nation's urge to exclude her from the post-conflict narrative. The violations by the nation made the world she lived an "unspeakable place, hostile and death-ridden" (Becker et al. 2000, 320).

Despite the unfriendly memoryscape, Terttu's story, as it is told in the documentary, becomes a success story. It is a story that is told in a learned and assertive voice. Her body is calm and she has found an inner peace with her memories. Her professional status as an interpreter and the choices made during her life are emphasized several times in her narrative. Terttu's body is calm and quiet compared to Elsa's vigorous and restless body. For Terttu, telling the narrative through an external language allows her memories become a part of a socially approved fabrication, i.e. a part of convention that defines what is believable. For her, the disciplinary measures imposed by the state have lost their power. The world is no longer an unfriendly place. She is no longer being named and there is a recollection and re-creation of events that now can find their ways to the post-conflict collective memory. Her truth, unlike Elma's, can be heard and is not ugly, for it responds to certain expectations of genre and structure. "A sequence in my life ended when I returned back to Finland" , declares Terttu. She renders her body memories tellable, which means that she orders and arranges them in the form of a chronological narrative. In Elma's case, on the other hand, the events themselves struggle to be told as they will. Her life does not have sequences since its time-frame follows a non-chronological order. For Terttu, her narrative places herself in the national history and, thereby, her life in the midst of others. She actively reclaims a sense of her whole self in order to confront the disciplinary practices that have cast her body out from the national self.

All three women in the documentary talk back to power. Their speech and bodies have no confessional tendencies since they are between two deaths, and closer to the second death their relation to power has started to erode. Elma, Terttu and Roosa are courageous without being confessional and hailing into being the agency of *superstite* as the young war veterans in the previous chapter (Chapter 4) did. Instead of offering a witness account what really happened as the Finnish Afghan war veterans presented in the chapter sought to do, these three women become *parrhesiastes*. *Parrhesiastes* is someone who presents an implicit or explicit diagnosis of the present in terms of power and knowledge. In short, the political act of *parrhesia* is a courageous act of truth-telling, of resisting power with truth. By manifesting to the community what they are, they reveal the subjectivity of the community. They do not confess what they have done, but problematize the politics that seeks closures in the form of historical truths. Stakes are high because the community does not want to hear the truth. However, they choose frankness and criticism instead of persuasion and confession.

The fact that a speaker says something different from what the majority believes is a strong indication that she is a *parrhesiastes*. When the speaker confronts the popular beliefs it builds anxiety and brings the trauma back. *Parrhesiastes* in this case disturbs the collective certainty by setting the peacebuilding record straight and offering a critical view of the memoryscape that tries to produce the fantasy of the coherent post-conflict collective self. Hence, the speaker speaks the truth ultimately by demonstrating the impossibility of the unitary post-conflict collective self (McGushing 2007; Foucault 2012).

Summary

In this chapter I have demonstrated how some bodies carry historical and political content that cannot be reconciled in the peacebuilding landscape. The bodies of women who fraternized with the German soldiers are examples of this type of post-conflict bodies. The Finnish national subject came into being through the establishment of "Hitler's brides" as others. The dangerous female body became a necessary signifier that had to be erased: the abject had to be killed, and, thereby, silenced in fantasy (cf. Corneytz 1999, 1–18). In Finland, the abject was "killed" and silenced through a variety of state-initiated disciplinary measures, including naming, medicalization, confession and non-recognition, which led to the expulsion the abject from the core of the nation.

Despite their vulnerability, post-conflict abject bodies can retain the right to remind the post-conflict collective self of the residual historical content that bears upon the present and cannot fully consume or dismiss its problematic past. Evidence that is corporeally sedimented can disturb the intensive attempts to produce the imaginary coherence during peacebuilding. The very materiality and the sheer presence of abject bodies in the midst of the collective self produced the disturbing effects. Furthermore, speaking back to the power coupled with corporeal performances are political and historical intervention: they question the linear peacebuilding path, offer a critical view of the memory work and move the collective self beyond its structural difference (cf. Guha 1999; Feldman 2004).

References

Aarnes, H. (2009). *Tyskertjentene: Historiene vi aldri ble foralt*. Oslo: Gyldendal.

Bakare-Yusuf, B. (1997). The Economy of Violence: Black Bodies and the Unspeakable Terror. In R. Lentin (Ed.), *Gender and Catastrophe* (pp. 171–185). London and New York: Zed Books.

Barad, K. (2012). On Touching—The Inhuman That Therefore I Am. *Differences: A Journal of Feminist Cultural Studies, 23*(3), 206–223.

Becker, G., Beyene, Y., & Ken, P. (2000). Memory, Trauma, and Embodied Distress: The Management of Disruption in the Stories of Cambodians in Exile. *Ethos, 28*(3), 320–345.

Berger, J. (1999). *After the End. Representations of Post-apocalypse*. Minneapolis: University of Minnesota Press.

Bleiker, R., & Hutchison, E. (2008). Fear No More: Emotions and World Politics. *Review of International Studies, 34*, 115–135.

Butalia, U. (1998). *The Other Side of Silence, Voices from the Partition of India*. New Delhi: Penguin Books.

Butler, J., & Spivak, G. (2007). *Who Sings the Nation-State? Language, Politics, Belonging*. New York and Calcutta: Seagull Books.

Caruth, C. (1995). Introduction. In C. Caruth (Ed.), *Trauma. Explorations in Memory* (pp. 151–157). Baltimore, MD: The Johns Hopkins University Press.

Clair, R. (1997). Organizing Silence: Silence as Voice and Voice as Silence in the Narrative Exploration of the Treaty of New Echota. *Western Journal of Communication, 61*(3), 315–337.

Cornyetz, N. (1999). *Dangerous Women, Deadly Words: Phallic Fantasy and Modernity in Three Japanese Writers*. Stanford: Stanford University Press.

Das, V. (1998). Language and Body: Transactions in the Construction of Pain. In A. Kleinman, V. Das, & M. Lock (Eds.), *Social Suffering* (pp. 67–92). Delhi: Oxford University Press.

Das, V. (2000). The Act of Witnessing: Violence, Poisonous Knowledge, and Subjectivity. In V. Das, A. Kleinman, M. Ramphele, & P. Reynolds (Eds.), *Violence and Subjectivity* (pp. 205–225). Berkeley, Los Angeles and London: University of California Press.

Das, V. (2007). *Life and Words. Violence and the Descent into the Ordinary*. Berkeley, Los Angeles and London: University of California Press.

Das, V. (2008). Violence, Gender, and Subjectivity. *Annual Review of Anthropology, 37*, 283–299.

Davidson, R., & Hall, L. (2001). Introduction. In R. Davidson & L. Hall (Eds.), *Sex, Sin and Suffering, Venereal Disease and European Society Since 1870* (pp. 1–5). London and New York: Routledge.
Delbo, C. (1990). *Days and Memory.* Marlboro, VT: The Marlboro Press.
Dollar, M. (2006). *A Voice and Nothing More* (R. Lamont, Trans.). Cambridge and London: The MIT Press.
Edkins, J. (2003). *Trauma and the Memory of Politics.* Cambridge: Cambridge University Press.
Edkins, J. (2006). Remembering Rationality: Trauma Time and Politics. In D. Bell (Ed.), *Memory, Trauma and World Politics* (pp. 99–115). New York: Palgrave Macmillan.
Ehlstain, J. B. (1987). *Women and War.* New York: Basic Books.
Enloe, C. (1990, September 25). Womenandchildren: Making Feminist Sense of the Persian Gulf War. *The Village Voice.*
Enloe, C. (2000). *Maneuvers. The International Politics of Militarizing Women's Lives.* Berkeley: University of California Press.
Ericsson, K. (2011a). Rape, Love and War—Personal or Political? *Theoretical Criminology, 15*(1), 67–82.
Ericsson, K. (2011b). "German Brats" or "War Children". In L. Westerlund (Ed.), *Saksalaisten ja neuvostosotilaiden lapset* (pp. 203–214). Helsinki: Nord Print.
Ericsson, K., & Simonsen, E. (2005). *Children of World War II. The Hidden Enemy Legacy.* Oxford and New York: Berg.
Feldman, A. (2004). Memory Theatres, Virtual Witnessing, and the Trauma-Aesthetic. *Biography, 27*(1), 163–202.
Fierke, K. (2004). Whereof We Can Speak, Thereof We Must Not Be Silent: Trauma, Political Solipsism and War. *Review of International Studies, 30*(4), 471–492.
Foucault, M. (1978). *The History of Sexuality Volume 1: An Introduction.* New York: Vintage Books.
Foucault, M. (2012). Parrhesiastes. Six Lectures by Michel Foucault at University of California, Berkeley, October–November 1983. Accessed 2 Apr 2012. http://72.52.202.216/~fenderse/Parrhesiastes.htm.
Gildea, R., Warring, A., & Wieveorka, O. (2006). *Surviving Hitler: Daily Life in Occupied Europe.* Oxford and New York: Berg Publishers.
Guha, R. (1999). The Small Voices of History. In S. Amin & D. Chakrabarty (Eds.), *Subaltern Studies IX, Writings on South Asian History and Society* (pp. 1–12). New Delhi: Oxford University Press.
Hayashi, H. (2008). Disputes in Japan Over the Japanese "Comfort Women" System and Its Perception in History. *The Annals of the Academy of the American Political and Social Science, 617*(1), 123–132.
Heiskanen, A. (2006). "Sitä taas eletään tätä päivää". Nuoren naisen elämä ja valinnat Kolmannen valtakunnan vaikutuspiirissä ja miehitetyssä Saksassa. In

T. Kinnunen & V. Kivimäki (Eds.), *Ihminen Sodassa* (pp. 261–274). Helsinki and Jyväskylä: Minerva.

Heiskanen, A. (2011). Kolmanteen valtakuntaan 1944 lähteneiden naisten kokemuksia äitiydestä ja selviytymisestä. In L. Westerlund (Ed.), *Saksalaisten ja neuvostosotilaiden lapset* (pp. 184–198). Helsinki: Nord Print.

Helén, I. (1997). *Äidin elämän politiikka*. Helsinki: Gaudeamus.

Hutchings, K. (2008). Making Sense of Masculinity and War. *Men and Masculinities, 10*(4), 389–404.

Jarvis, C. (2004). *The Male Body at War. American Masculinity During World War II*. Northern Illinois: Northern Illinois University Press.

Junila, M. (2000). *Kotirintaman aseveljeyttä*. Helsinki: Suomalaisen Kirjallisuuden Seura.

Kleinman, A., & Kleinman, J. (1994). How Bodies Remember: Social Memory and Bodily Experience of Criticism, Resistance, and Delegitimation Following China's Cultural Revolution. *New Literary History, 24*, 707–723.

Kontinen, Ville. (2011). Lemmestä, pelosta ja pakostakin. Suomalaisten motiivit saksalaisjoukkojen mukaan lähtemiselle syksyllä 1944. In L. Westerlund (Ed.), *Saksalaisten ja neuvostosotilaiden lapset* (pp. 159–179). Helsinki: Nord Print.

Kundrus, B. (2005). Forbidden Company: Romantic Relationships Between Germans and Foreigners, 1939 to 1945. In D. Herzog (Ed.), *Sexuality and German Fascism* (pp. 201–222). New York and Oxford: Berghahn Books.

Leys, R. (2000). *Trauma: A Genealogy*. Chicago: Chicago University Press.

Lock, M., & Kaufert, P. (1998). Introduction. In M. Lock & P. Kaufert (Ed.), *Pragmatic Women and Body Politics* (pp. 1–27). Cambridge: Cambridge University Press.

Markkanen, K. (2011). "Hitler's Brides" from Finnish Lapland. *Helsingin Sanomat, International edition*. Accessed 30 Mar 2012. http://www.hs.fi/english/article/%E2%80%9CHitler%E2%80%99s+brides%E2%80%9D+-from+Finnish+Lapland/1135266026489.

McGushing, E. (2007). *Foucault's Askesis. An Introduction to the Philosophical Life*. Evanston: Northwestern University Press.

Mosse, G. (1996). *The Image of Man: The Creation of Modern Masculinity*. Oxford: Oxford University Press.

Motsemme, N. (2004). The Mute Always Speak: On Women's Silence at the Truth and Reconciliation Commission. *Current Sociology, 52*(5), 909–932.

Nagel, J. (1998). Masculinity and Nationalism. *Ethnic and Racial Studies, 21*(2), 242–269.

Parashar, S. (2011). Women Militants as Gendered Political Subjects. In A. Kronsell & E. Svedberg (Eds.), *Making Gender, Making War* (pp. 299–325). New York: Routledge.

Radstone, S. (2007). Trauma Theory: Contexts, Politics, Ethics. *Paragraph, 30*(1), 9–29.

Riessman, C. (2003). Women and Medicalization. A New Perspective. In R. Weitz (Ed.), *The Politics of Women's Bodies* (pp. 46–63). Oxford and New York: Oxford University Press.

Ross, F. (2001). Speech and Silence: Women's Testimony in the First Five Weeks of Public Hearings of the South African Truth and Reconciliation Commission. In V. Das, A. Kleinman, M. Lock, M. Ramphele, & P. Reynolds (Eds.), *Remaking a World Violence, Social Suffering, and Recovery* (pp. 250–280). Berkeley: University of California Press.

Sauerteig, L. (2002). The Fatherland Is in Danger, Save the Fatherland. In R. Davidson & L. Hall (Eds.), *Sex, Sin and Suffering, Venereal Disease and European Society Since 1870* (pp. 76–92). London and New York: Routledge.

Scarry, E. (1985). *The Body in Pain. The Making and Unmaking of the World.* Oxford: Oxford University Press.

Shapiro, M. (1997). *Violent Cartographies. Mapping Cultures of War.* Minneapolis: University of Minnesota Press.

Sjoberg, L., & Gentry, C. (2007). *Mothers, Monsters, Whores: Women's Violence in Global Politics.* London: Zed Books.

Soh, S. (2008). *The Comfort Women. Sexual Violence and Postcolonial Memory in Korea and Japan.* Chicago and London: The University of Chicago Press.

Stoller, P. (1994). Embodying Colonial Memories. *American Anthropologist, 96*(3), 634–648.

Tanno, D. (1994). Names Narratives and the Evolution of Ethnic Identity. In A. Gonzalez, M. Houston, & V. Chen (Eds.), *Our Voices: Essays in Culture, Ethnicity, and Communication* (pp. 38–41). Los Angeles, CA: Roxbury Press.

Timm, A. (2005). Sex with a Purpose: Prostitution, Venereal Disease, and Militarized Masculinity in the Third Reich. In D. Herzog (Ed.), *Sexuality and German Fascism* (pp. 223–255). New York and Oxford: Berghahn Books.

Todeschini, M. (2002). The Bomb's Womb? Women and the A-Bomb. In V. Das, A. Kleinman, M. Lock, M. Ramphele, & P. Reynolds (Eds.), *Remaking a World Violence, Social Suffering, and Recovery* (pp. 102–156). Berkeley: University of California Press.

Uhlenius, P. (2011). Saksalaissotilaiden ja neuvostovankien salatut lapset. In L. Westerlund (Ed.), *Saksalaisten ja neuvostosotilaiden lapset* (pp. 121–152). Helsinki: Nord Print.

Väyrynen, T. (2013). Keeping the Trauma of War Open in the Male Body—Resisting the Hegemonic Forms of Masculinity and Finnish National Identity. *Journal for Gender Studies, 22*(1), 1–15.

Virgili, F. (2002). *Shorn Women: Gender and Punishment in Liberation France.* Oxford and New York: Berg Publishers.

Virolainen, K. (1999). *Elinikäinen taakka.* Rovaniemi: Lapin yliopisto.

Westerlund, L. (2011). *Saksalaisten sotilaitten lapset.* Helsinki: Nord Print.

Yuval-Davis, N. (1997). *Gender and Nation.* London: Sage.

Yuval-Davis, N., & Anthias, F. (Eds.). (1989). *Women—Nation—State.* Basingstoke: Macmillan.

Žižek, S. (2006). Series Foreword. In M. Dollar (Ed.), *A Voice and Nothing More* (pp. i–ii). Cambridge and London: The MIT Press.

CHAPTER 6

Peacebuilding as a Corporeal, Temporal and Mneumonic Site

Abstract Peacebuilding is marked by heterogenious histories and tense locations of political differences. How the past conflict is remembered is fundamental to peacebuilding as the practices of remembering war and violence bring forth the corporeal politics of peacebuilding. The past conflict is not just remembered, but it is enacted in bodies, i.e. corporeally remembered. During peacebuilding the stakes are high since it always involves a struggle who has the right to present post-conflict community, who has right to be visible and, in more general, who belongs to the community. This chapter discusses two remembering, yet resistant bodies through Jacques Rancière's notion of the political. The political is understood her to be a struggle of logics that count the parties and parts of the post-conflict community.

Keywords Reconciliation · Memory · Post-conflict communities

Peacebuilding is marked by heterogenious histories and tense locations of political differences. How the past conflict is remembered is fundamental to peacebuilding as the practices of remembering war and violence bring forth the corporeal politics in peacebuilding as the past conflict is not just remembered, but it is enacted in bodies, i.e. corporeally remembered. I examine in this chapter how peacebuilding is a temporal, corporeal and mneumonic site that is characterized by an intensive

T. Väyrynen, *Corporeal Peacebuilding*, Rethinking Peace and Conflict Studies, https://doi.org/10.1007/978-3-319-97259-6_6

struggle how the war and violence are remembered, what is allowed to be remembered and how memory is transferred across time, namely how the temporal and corporeal transition of conflict to peace takes place. Ultimately, the question is about bodies that are transformed by political processes and bodies that experience the processes as lived memory. In order to explore the problematique, I introduce a Bergsonian ontology of becoming where the world is a site of endless and unchartable becomings and where the real is "understood as durational: it is composed of millions even billions of specific durations, each with its own measure, its own span" (Grosz 2005, 11).

I argue that the struggle for memory, or rather memory work, is vital for the re-establishment of the post-conflict power. As discussed in the previous chapter (Chapter 5), there is an acute quest for a consensual view of the past during peacebuilding in order to produce a coherent and unitary post-conflict collective self. The quest, however, is never completed as post-conflict collective memory work is an ongoing political project which aims at realigning memory with collective identity. The project is complex as memory work is not just an intellectual project that takes place in one site and at a certain point of time, but it includes corporeal transition of the past to the present. I explore in this chapter the importance and unexpectedness of the mundane and corporeal in relation to the long-term transition from war to peace, namely long-term peacebuilding. Peacebuilding is understood to be in this book an ongoing (political) process with multiple temporal orders and durations. In order to examine the corporeal mneumonics I employ two post-conflict narratives, namely Kaisu's and Seppo's stories. The source material is, however, not just their stories, but their corporeal performances that are ultimately about the political of peacebuilding, namely how the past violence is remembered and who belongs to the post-conflict political community.

In Kaisu's case the political performance took place when she appeared in the document (Suutari 2010) discussed in Chapter 5 and in a variety of media interviews (YLE 2010; Nykänen 2010) after the documentary had been released. In Seppo's case the question is about his critical engagement with war, male body and remembering through art. Their experiences of war and violence are indeed different. Kaisu is one of the Finnish women who left with the withdrawing German army and then returned back to Finland, whereas Seppo's father fought in the Second World War. Seppo is a son of a war veteran, whereas Kaisu experienced the war herself. War and violence touched them both and rendered them vulnerable in multiple ways. By using this material I seek to demonstrate how memories of war and violence are embedded in bodies and how also remembering bodies transfer war to peacebuilding and peace. Through Kaisu's and Seppo's corporeal mneumonics it is possible, in my view, to examine long-term peacebuilding and the political that emerges when intensive and painful memories of war, conflict and violence are brought fore and made visible in peacetime. The chapter hence attests to a "liminal signifying space" (Bhabha 1994, 148) that always characterizes peacebuilding.

As indicated, this chapter returns to the documentary film by Virpi Suutari titled *Auf Wiedersehen Finnland*. In addition to the film's visual material, I employ less used material in Peace and Conflict Studies, namely art as I analyze one of Salminen's works, a performance titled *The Miracle of Snow* (originally performed in 1991). I consider art to important "data " for critical Peace and Conflict Studies since it can challenge the taken-for-granted facts about the past. As Jill Bennett (2005, 11) writes, art carries an ability to "thrust us involuntarily into a mode of critical inquiry" by challenging, for example, conventional identity narratives. Even more importantly, when the focus is on the contested site of peacebuilding and since art often plays with ambiguity, artistic acts can "intervene in the constitution of the political by reshaping what is visible, sayable and doable" as David Ingram (2012, 131) argues about the capacity of art to evoke the political. In other words, art reveals the limits what can be visible and who can belong to the post-conflict political community. By doing this, art functions itself as a critical and even disruptive force.

What becomes visible, sayable and doable, is a matter of politics as indicated in the earlier chapters. To study the political in peacebuilding, this chapter derives again from Jacque Rancière's understanding of politics that is characterized by the de-classification of spatial and temporal orders as a precondition for politics. From this position, when speech or visibility is acquired by those whose right to speak or to be visible is not recognized they can produce spatiality and temporality that disturbs hegemonic representations of post-conflict time-space constellations, namely they force the community to ask "who are we?" and "where are we?" (Rancière 1999, 99–100; 2009, 116). In the previous chapter (Chapter 5) women took the position of *parrhesiastes* in relation to the collective self. Here, on the other hand, the type of peacebuilding politics which emerges from Kaisu's and Seppo's performances highlights that knowing who is qualified to become visible and how one is qualified to become visible is at the core of the political since "politics revolves around the properties of spaces and the possibilities of time" (Rancière 2003, 201).

Body Time

There is no singular temporality in peacebuilding that characterizes the long-term process and the corporal and temporal transitions that are an incremental part of the process. Rather, peacebuilding time is "time multiple". In addition with the linear national time (which will be discussed in more detail in Chapter 7) there is body time on which this chapter focuses. Christian Haine (2011, 116) writes with reference to Deleuze, that there is "body time" which is "finite in that it manifests itself in particular modes of activity, but infinite in that it operates through a process of relation that is open, composing itself through contingent, inventive combinations". Kaisu's performance exemplifies body time as it consists of layers of time that do not coincide with the linear peacebuilding national time. Seppo's time, on the other hand, connects his veteran father's body to his his own living body and, thereby, introduces a break to the post-conflict national narrative and its unitary time.

Kaisu is an elegant grey haired slim who speaks in a calm voise. She says that she left Finland with the Nazi forces in 1944 because she was afraid of a Soviet occupation. She did not have an affair with a German soldier, but rather saw the possibility of employment with the withdrawing German troops and, at the same time, escape the Soviet threat. "I escaped the Russkie" and "was given the identity of Gretel Laubert for a short while during my escape", she says in the documentary. Kaisu gives detailed accounts of how she traveled through Norway with the withdrawing German troops to Germany, held a variety of occupations there and saw the collapse of the Nazi regime and the total destruction of the German cities.

Kaisu reminisces how "most of the Finns thought that the conditions in Germany would be better, but to our surprise, that was not the case". She remembers how "Hitler proclaimed that the Finnish girls can be treated as sisters-in-arms and can enter Germany". She spent several months in Germany and during the last months of the war in 1945 she was transferred to a transit camp from where she escaped back to

Finland. In Finland, she was incarcerated in a detention centre administered by the Finnish security police. She was interrogated there several times. Her voice fails three times in the documentary: when she apologies for the hardship she has caused her family, when she remembers the suffering of the Jewish population and when she tells about her humiliation at the detention centre.

Kaisu's narrative shows how there are loops and tangles in post-conflict memory as time easily looses its linear qualities in relation to conflict and violence. Narrative loops and tangles indicate in Kaisu's case that her temporality does not coincide with the peacebuilding linear national temporality and the selective and constrained memoryscape enabled by it. "Do excuse me the suffering I have caused to you, but there was nothing else I could have done" reads Kaisu from a letter she has written to her relatives on the 15th of September 1944 where she explains her reason for leaving with the withdrawing German army. When Kaisu reads a letter in the documentary that she has written almost seventy years ago, her time slows down and, as Henri Bergson (1910, 100) writes about pure duration, her time "refrains from separating its present from its former states". Time loses its measurable and linear qualities when the multiple layers of possible pasts actualize in the act of reading the letter. Her reading of it consists of the memory of the event of writing the letter which, on the other hand, contains other memories that can be interpreted from her corporeal components as her body re-enacts the moment of writing, her mother's reading of the letter and the mother's reaction to it. The reading of the letter in the present moment blurs the distinction between the past and the present moment as it brings the past of war to the present of peace where they form an organic whole. Her temporality is incomplete and continuously growing as there is no beginning nor ending to it.

As indicated earlier, corporeal temporalities do not follow the assumed linear governmental logic of peacebuilding where war terminates in peace and where these two can be clearly distinguished from each other within a temporal frame that distinguishes the past from the present. Henri Bergson's notion of time as a whole, *durée*, captures the "other temporalities", namely individual and corporeal temporalities that are equally important for peacebuilding. Time is not, in the Bergsonian view, a linear and causal line of the transition—for example from war to peace through peacebuilding—where past precedes present and is succeeded by future, but a 'surface' where past and present exist symbiotically together and are in a constant process of bodily becoming. Moreover, duration disturbs the linear and causal timeframe by setting the temporalities in direct confrontation where the future does not follow the present and past. For example, when Kaisu acknowledges her violent past in the present, the fusion of the past, present and future follows the acknowledgement. The past is not based simply on Kaisu's direct experience, but it is a mosaic of what she knew first hand, what she was told, what she imagined, and what happened around her as part of a historical process that she shared with millions of other people. Many pasts exist in parallel and some of them have been previously wiped out due to their disturbing qualities. In this temporality, the memory of the past can make the past out as not necessarily true, or the past can be misrecognized, as the past and present exist on the same surface, not in linearity, but repeatedly in flux, flowing from one to another (Bergson 1911; Deleuze 2005; Huchings 2011).

Relational Bodies and Splitting Temporalities

War and violence do not shun away bodies' capacity to empathetically relate to the alien, even enemy, bodies and to resonate with their vulnerability and suffering. As discussed in Chapter 3, local women reached towards the Russian prisoners of war during the Second World War despite that interaction with them was strictly forbidden by the Finnish state. Giving bread to the prisoners was an act of recognition and acknowledgement and it rendered the prisoners' bodies visible for the whole community. In a similar manner, Kaisu relates to the Jewish population in Germany in the documentary film when she remembers their suffering. She describes in detail how the Jews were contained in small underground bunkers and how their bodies were malformed due to that. "I and some other Finnish girls encountered a Jewish woman who wore a yellow patch with the Jewish star. Her eyes were full of anguish and they were begging us to notice that she was indeed wearing the patch". She wonders what kind of suffering has caused such timidity in a fellow human being. Her voice becomes low and husky when she reminisces the suffering of the others. Her own hardship in war-torn and ravaged Germany seems to have only minor importance compared to the suffering and vulnerability of the Jewish population.

When Kaisu mourns the suffering of the others, a splitting of the subject and temporality takes place. She is not Kaisu located here and now, but young Kaisu who glimpsed into the void of human existence that is relational and always already vulnerable. She loses her notion of herself as an autonomous subject and in control. Judith Butler (2004, 23) describes the event of mourning the other:

> What grief displays is the thrall in which our relations with others holds us, in ways that we cannot always recount or explain, in ways that often interrupt the self-conscious account of ourselves we might try to provide, in ways that challenge the very notion of ourselves as autonomous and in control. I might try to tell a story here, about what I am feeling, but it would have to be a story in which the very "I" who seeks to tell the story is stopped in the midst of the telling; the very "I" is called into question by its relation to the Other, a relation that does not precisely reduce me to speechlessness, but does nevertheless clutter my speech with signs of its undoing. I tell a story about the relations I choose, only to expose, somewhere along the way, the way I am gripped and undone by these very relations. My narrative falters, as it must.

When the autonomous "I" rambles, the chronological timeframe loses its grip. In this splitting, the "subject is at once both inside and outside, self and other, virtual actual, recollection and perception and, indeed, past and present" (Martin-Jones 2008, 30). Kaisu's temporality is without a centre and it is expanding as is enlarges outwards towards the suffering of the others. Her temporality becomes a series of infinitely bifurcating pathways. Kaisu herself becomes a discontinuous corporeal entity who simultaneously exists in multiple temporalities. Here her time moves backwards in a non-linear way to uncover the other pasts and the pasts of the others.

Peacebuilding does not remove the existence of the variety of violence. In one of the most affective scenes of the documentary, Kaisu sits on the stairs of a grey wooden hut where she was incarcerated by the security police on her arrival back in Finland. She smokes a cigarette with a firm hand, but the viewer can see how her overtly upright body holds the memories of the past where the representatives of the Finnish nation interviewed her several times in the most humiliating ways. "Hautojärvi – that was his name – took me three times to special interrogations where no notes were taken, and I can tell you, he was brute" says Kaisu and continues that "however, in front of Hautojärvi I did not cry!". That particular layer of the past and its temporality is mainly beyond language and it is incorporated into her body in the form of pain and grief. For Kaisu, "what is 'remembered' in the body is well remembered", as Elaine Scarry (1985, 109) argues when she discusses torture.

The codified post-conflict national memory skates along the flat surface of the past, whereas the past that is remembered in the body penetrates the deepest layers of human existence as Kaisu's peacetime performance indicates. Kaisu's body has indeed been inscribed by a variety of disciplinary practices as it has been located into the national order of things. The sovereign power has named her, medicalized her body and subjected her to the mechanisms of surveillance as described in the previous chapter (Chapter 5). Chronological national time has, therefore, taken hold of her body. Yet, Kaisu's temporality writes itself out of normalized post-conflict national time. The body is transformed by processes and does not only represent those processes, but experiences them as the lived memory constantly in flux and in the process of becoming (Becker et al. 2000; Connerton 1989; Culbertson 1995; Das 1998; Dossa 2003; McLane 1996; Sipllman and Conway 2007). The time of Kaisu's body is in this sense immeasurable.

Corporeal Transference of War Memory

As indicated earlier, violence does not end with peacebuilding. Rather, it gets new meanings through and in the bodies of future generations. Artist Seppo Salminen returns to war time violence in his performance *The Miracle of Snow: On Father's Grave* (1991). In the performance he stands barefoot on a black steel platform covered with two inches of frost and cuts symbolically his wrists open. His saggy shirt and the white layer of frost are soon marked with physical stains of blue blood, and the artist falls down to embrace the smeared whiteness. Seppo's own body seems to become one with the icy grave in which his father rests.

Seppo's symbolic self-sacrifice in his father's grave can be interpreted to explore the transference of violence from war to peace through corporeal mneumonics. Unlike Kaisu, Seppo did not experience war himself, yet his body remembers war and violence. Seppo indicates in his exhibition catalogue that his father was a Second World War veteran and his father's wounded body had been a formative influence in his life. He realized very early in his childhood that there is an intimate connection between the male body and the identity of the (post-conflict) nation. Salminen states in the catalogue that "it's [the male body] not his own, it belongs to a nation or to whomever is executing its sacrifice" and "the male body becomes sacred only through sacrifice" (Erkkilä 2008, 3–4).

Seppo Salminen concurs with Josh Tosh (2004, 49) who argues that powerful and invulnerable male body invokes ideals of sacrifice, and in times of national crisis "this dominant masculinity is likely to become a metaphor for the political community as a whole and to be expressed in highly idealized forms". During wartime, nationalism flourishes and bodies are categorized and ranked into those who are capable of defending and securing the nation or the community (e.g. heroic warriors, sacrificial mothers) and those who are a risk to the collective identity (e.g. conscious objectors, women who fraternize with enemy soldiers), and those categorizations and differences are extended into peacebuilding too. Whilst war is not an isolated political, social and cultural institution, it shapes ideals of manhood, as well as ideal corporeal being, beyond the war. In Michael Shapiro's (2002, 258) words, war serves to impose a "coherence on the citizen's body" and obedient nationalistic bodies are created by forging the ties between (nationalistic) ideology and the body in war which is carried over to peacetime.

Walking barefoot in icy snow is extremely painful, almost impossible. In cutting open his wrists, Salminen can be seen to act out the destiny of the male body, which is to be sacrificed in the name of the nation and its fantasized war miracles. He (Salminen 2008) makes the connection between the nation and the male body whose destiny is to become sacrificed:

> I felt that my father had been a victim of history. Like many other men of his generation, he ended up in the war and understood that the male body is a site of execution. The purpose of a man's body is to become mutilated.

Idealized male (and female) bodies play an important role in stitching together the post-war collective identity. As indicated in Chapter 5, literature on gender and nationalism has demonstrated how nationalistic rhetoric, be it verbal or visual, validates male bodies primarily by reference to men's supreme sacrifice for their nation. This literature establishes connections between gender, the body and the nation by arguing that the nation takes hold of the human body and produces gendered differences that symbolically represent the nation. The literature concurs with the view that the human body can be seen to be molded by a variety of disciplinary practices. It is argued in this gender-nationalism nexus literature, that male bodies that are defending the nation, freedom and honor are encoded so as to embody the ultimate manly virtues of courage, discipline, competitiveness, virility, dignity and strength.

In her influential book *The Body in Pain* (1985), Elaine Scarry takes a view beyond war and argues that the nation penetrates the deepest layers of consciousness and manifests itself in the body. In a similar vein, Veena Das (2008, 290) writes about empires that "individuals are interpellated in the structure of empire so that even when they are not themselves dominant within patriarchal and racistic hierarchies, they can claim inclusion with the projects of empire by literally embodying it". According to Scarry and Das, there is a dialectic between *habitus* and objective social structures as the source of social incorporations into the body. As already indicated in the previous chapter (Chapter 5), it is argued in this tradition of thought that bodies are transformed by political processes, such as conflict and peacebuilding, and bodies experience the processes as the lived memory.

Instead of taking part in the national fantasy of the strong male body, Salminen's barefoot performance *The Miracle of Snow* can be interpreted as a challenge to the standardized ways of narrating conflict history by reference to inyealding male body and its sacrificial qualities. Rather, the performance can be read to indicate the vulnerability of the male body. Salminen's performance establishes a relation between his self-sacrificial and vulnerable body, his veteran father's wounded body and the sacrifices that were assumed from the Finnish soldiers during the Second World War. By doing this the body produces a temporal and corporeal transition from war to peace. In other words, the performance can be seen to examine they ways war is transferred in bodies across time to future generations and how bodies are always already vulnerable despite the imposition of the idealized ways of being a part of post-war collectivity. Salminen's father's body had been subjugated to the intensified control of bodies due to the war and its regime, whereas Salminen's body and its temporality carry the "record of war" (Scarry 1985, 113) and it transfers the past violence to present peace.

As argued on Chapter 3, Marianne Hirsch's (2008) concept of "postmemory" captures particularly the second-generation traumatic and embodied knowledge. Postmemory is, according to Hirsch, generated when events happened in the past but their effects continue in the present. It is corporeal in the sense that it is transmitted through the language of the body. By being embodied memory and being transferred through bodies, i.e. so-called habit memory, it differs radically from the national archival memory that is transferred largely through language. The post-war national archival memory seeks closures that provide the image of the unitary national self whereas for this type of habit memory there is no closure.

Summary

Kaisu's, as well as the other women who left with withdrawing the German army discussed in Chapter 5, body is constituted outside and it is unaccounted for in the symbolic order of the post-conflict nation. Seppo's performance, on the other hand, critically engages with the institution of war and explores how war is transferred through bodies to peace. Their performances open up the possibility, if not the necessity, for the enactment of the political as they compose themselves through contingent, inventive and interruptive combinations when making visible the corporeal mneumonics of peacebuilding.

The transition their bodies make is ultimately about politics as the heterogenious histories and the tense locations of political differences emerge in them. As Rancière (2001, thesis 6) notes, political struggle is not a conflict between well-defined interest groups; it is an opposition of logics that count the parties and parts of the community in different ways. He (Rancière 2001, thesis 7) writes:

> The essence of politics, then, is to disturb this arrangement by supplementing it with a part of the no-part identified with the community as a whole. Political litigiousness/struggle is that which brings politics into being by separating it from the police that is, in turn, always attempting its disappearance either by crudely denying it, or by subsuming that logic to its own. Politics is first and foremost an intervention upon the visible and the sayable.

Politics in Kaisu's and Seppo's cases is signaled by the enmeshing of the corporeal temporalities that create a moment of interruption, a moment where their performances disturbs the post-war national order of thing, its imagined temporal coherence and its idealized forms of belonging to the community. It is not their speaking and the content of what they say, but the corporeal post-conflict memory and temporality through which they make an intervention upon visible. The intervention disturbs the singular form of the national post-war linear narrative by introducing the political potential embedded in multiple parallel temporal universes. They offer alternative histories without them being specifically given as correct. The fusing of the post-war national "true" past with their pasts cause the resurgence of difference and contestation that destabilizes the already established national, teleologically determined, transition from war to peace.

It is in this sense that Kaisu and Seppo introduces political struggle. They do not rupture the power positions, but rather challenge the classifications of those worthy of inclusion and those excluded as well as bringing forth a space for the appearance of corporeal mneumonics. Kaisu enables herself and her kind to be seen and heard as speaking subjects and engages in a radical political practice that displaces the limits of post-conflict social exclusions. Seppo, on the other hand, demonstrates how the idealized male bodies hailed into being during war, haunt future bodies in the form of corporeal postmemory. His body is vulnerable and thereby is does not fit into the model of inyelding male citizen. Political struggle in this sense is not a rational debate between multiple interests, as noted earlier, but the struggle for one's voice to be recognized as the voice of a legitimate partner and one's corporeal ways of remembering as a legitimate way of making a transition from war to peace.

References

Bennett, J. (2005). *Empathic Vision: Affect, Trauma, and Contemporary Art.* Stanford: Stanford University Press.

Bergson, H. (1910). *Time and Free Will: An Essay on the Immediate Data of Consciousness* (F. L. Pogson, Trans.). London: Allen and Unwin.

Bergson, H. (1911). *Matter and Memory.* London: Swan Sonnenschein.

Becker, G., Beyene, Y., & Ken, P. (2000). Memory, Trauma, and Embodied Distress: The Management of Disruption in the Stories of Cambodians in Exile. *Ethos, 28*(3), 320–345.

Bhabha, H. (1994). Dissemination: Time, Narrative, and the Margins of the Modern Nation. In H. Bhabha (Ed.), *Nation and Narration* (pp. 139–170). New York: Routledge.

Butler, J. (2004). *Precarious Life: The Powers of Mourning and Violence.* London: Verso.

Culbertson, R. (1995). Embodied Memory, Transcendence, and Telling: Recounting Trauma, Re-establishing the Self. *New Literary History, 26*, 169–195.

Connerton, P. (1989). *How Societies Remember*. Cambridge: Cambridge University Press.

Das, V. (1998). Language and Body: Transactions in the Construction of Pain. In A. Kleinman, V. Das, & M. Lock (Eds.), *Social Suffering* (pp. 67–92). Delhi: Oxford University Press.

Das, V. (2008). Violence, Gender, and Subjectivity. *Annual Review of Anthropology, 37*, 283–299.

Deleuze, G. (2005). *Cinema 2: The Time Image*. London: Continuum.

Dossa, P. (2003). The Body Remembers: A Migratory Tale of Social Suffering and Witnessing. *International Journal of Mental Health, 3*(3), 50–73.

Erkkilä, H. (2008). Johannes s: isän ja pojan trauman näyttämö – psykoanalyyttinen näkökulma. In *Johannes s Still Life* (Exhibition Catalogue). Helsinki: Nykypaino.

Grosz, E. (2005). Bergson, Deleuze and the Becoming of Unbecoming. *Parallax, 11*(11), 4–13.

Haines, C. (2011). Corporeal Time: The Cinematic Bodies of Arthur Rimbaud and Gilles Deleuze. *Angelaki: Journal of the Theoretical Humanities, 16*(2), 103–126.

Hirsch, M. (2008). The Generation of Postmemory. *Poetics Today, 29*(1), 103–128.

Hutchings, K. (2011). What Is Orientation in Thinking?: On the Question of Time and Timeliness in Cosmopolitical Thought. *Constellations, 18*(2), 190–204.

Ingram, A. (2012). Experimental Geopolitics: Wafaa Bilal's Domestic Tension. *The Geographic Journal, 178*(2), 123–133.

Martin-Jones, D. (2008). *Deleuze, Cinema and National Identity*. Edinburgh: Edinburgh University Press.

McLane, J. (1996). The Voice on the Skin: Self Mutilation and Merleau Ponty's Theory of Language. *Hypatia, 11*(4), 107–118.

Nykänen, A.-S. (2010, March 21). Saksalaisten matkaan 1944 lähteneet naiset epäröivät yhä puhua. *Helsingin Sanomat*.

Rancière, J. (1999). *Disagreement*. Minneapolis: University of Minnesota Press.

Rancière, J. (2001). Ten Theses on Politics. *Theory and Event, 5*(3). http://www.egs.edu/faculty/jacques-ranciere/articles/ten-thesis-on-politics/. Accessed 1 Oct 2014.

Rancière, J. (2003). Politics and Aesthetics: An Interview. *Angelaki: Journal of the Theoretical Humanities, 8*(2), 119–211.

Rancière, J. (2009). A Few Remarks on the Method of Jacques Rancière. *Parallax, 15*(3), 114–123.

Salminen, S. (originally performed in 1991). *The Miracle of Snow: On Father's Grave*. Helsinki.

Salminen, S. (2008, November 11). Interview. *Helsingin Sanomat*.

Scarry, E. (1985). *The Body in Pain: The Making and Unmaking of the World*. New York and Oxford: Oxford University Press.

Shapiro, M. (2002). Partition Blues. *Alternatives, 27*(2), 249–271.

Spillman, L., & Conway, B. (2007). Texts, Bodies, and the Memory of Bloody Sunday. *Symbolic Interaction, 30*(1), 79–103.

Suutari, V. (2010). *Auf Wiedersehen Finnland*. Helsinki: ForReal Productions.

Tosh, J. (2004). Hegemonic Masculinity and the History of Gender. In S. Dudink, K. Hagemann, & J. Tosh (Eds.), *Masculinities in Politics and War: Gendering Modern History* (pp. 41–60). Manchester: Manchester University Press.

YLE. (2010, January 28). *Yle aamu-tv*. http://yle.fi/uutiset/saksalaissotilaiden_matkaan_lahteneet_avaavat_kipeita_muistojaan/5498851. Accessed 10 May 2014.

CHAPTER 7

Peacebuilding in Colonial Relations

Abstract Colonial relations are seldom discussed in terms of peacebuilding. Yet, dismantling colonial relations requires comprehensive processes, approaches and stages that transform the relations toward more sustainable, peaceful, relations. Nations do not always want to recognize colonial relations and the existence of the indigenous populations at the core of the nation and, thereby, long-term peacbuilding is impossible. This chapter discusses the nexus of colonial violence and peacebuilding through the example of Finnish-Sámi settler colonial relations. Violence that functions though the indigenous bodies render the bodies mute as violence embedded in these relations is structural, slow as well as epistemic. The chapter continues the discussion on the preconditions of the political and peacebuilding started in the previous chapter.

Keywords Peacebuilding · Post-colonial theory · Indigenous activism

Dismantling colonial relations requires comprehensive processes, approaches and stages that transform the relations toward more sustainable, peaceful, relations. Yet, colonial relations are seldom discussed in terms of peacebuilding. In this chapter I seek to fulfill this gap in Peace and Conflict Studies by outlining that transforming colonial violence is not merely a stage in time or a condition, but a dynamic long-term social construct (cf. Lederach 1997, 20)—it is, hence, peacebuilding. I see colonialism to be an entanglement of political social and economic

T. Väyrynen, *Corporeal Peacebuilding*, Rethinking Peace and Conflict Studies, https://doi.org/10.1007/978-3-319-97259-6_7

practices, ideologies, relationships, collections and archives since colonialism is not just about land and resources, but it is about bodies (cf. de Leeuw 2016, 14). In other words, colonial violence operates through bodies, and it includes multiple forms of violence. While operating through bodies, colonialism is always multiple and intersectional as there are "vastly different colonial experiences, for both colonized and colonizer, structured through constructions of race, class and gender" (Edwards and Mead 2013, 20). In this chapter I focus on the experiences of the colonized and the preconditions for peacebuilding.

This chapter is about corporeal peacebuilding in relations where there has not been overt violence. I use material from the indigenous Sámi-Finnish relations that fulfil many of the characteristics that are constitutive of colonial relations in general, and settler colonial relations in particular. I concur with the Sámi activists who argue that the Finnish nation has repressed and co-opted the indigenous Sámi alterities in the name of unitary nation. The violence embedded in these relations has gone unrecognized since it is structural as well as epistemic in the sense that it functions through corporeal practices that seek to neutralize differences, produce standardized patterns of being a citizen and render alternative forms of existence, histories and futures unrecognized. However, identifying the variety of long-term practices that do harm and violence is often difficult exercise as colonialism establishes a visceral realm where social difference are operating in the "body alongside other processes more often recognised as cellular, molecular or bodily" (Hayes-Conroy and Martin 2010, 272). In this chapter I also attend to the importance of visceral realm and differences created through it.

In Finland, the official national history has not recognized the colonial nature of the Sámi-Finnish relations. There is, however, a new collective urge to re-write this part of Finnish history. In summer 2017 when Finland celebrated its 100 years of independence, some influential Finnish historians pointed out that the silenced Sámi-Finnish settler colonial relations are among some of the worst atrocities that have taken place during the hundred years of Finnish independence. They argued that the treatment of the Sámi population forms a black spot in the archival history writing that seeks to establish and maintain an image of the unitary nation-state and its non-colonial history. For example, professor Marjo Kaartinen noted that the Sámi identity was stripped away in the 1960s by forcing the Sámi population to use Finnish language. In a similar vein, professor Juhani Suomi recognized that Finland has not cratified the ILO 169 treaty on the rights of the indegenious and tribal populations (Helsingin Sanomat 6.8.2017). Finland, however, has wanted to represent itself as a peaceful country without colonial history although the violent settler colonial relations are at the core of the nation.

The corporeal practices of governance through which the Sámi population is controlled include Christianization, taxation, schooling, appropriaton of a genedered bodies for a variety of purposes and, ultimately, assimilation (Spangen et al. 2015, 1). The practices establish long-term patterns of repression which is not a single violent event with an endpoint, but rather it is based on constantly—and slowly—operating practices that have structural qualities (cf. Veracini 2013). Finland has granted the Sámi population a limited right to govern through the Sámi Parliament, but as Glen Coulthard (2007) argues, settler polities may grant indigenous nations this type of recognition precise to contain them as a domesticated difference that facilitates state authority over land and the definition of indigenous identity. Moreover, the difference is naturalized not only in social and political spaces, but also in definitions and experiences of subjectivity (Morgensen 2011, 16). Given this domesticated difference that is foundational for many colonial relations, I show in this chapter that its denaturalization is a prerequisite for long-term corporeal peacebuilding in colonial relations.

Transformative Potential of Art

As demonstrated in Chapter 6 in reference to Jill Bennett's (2005) work, art can contribute to peacebuilding due to its critical and transformative potential. In short, art can have a double function: it can consist of collective affective potential for transformation and form a critical inquiry itself. In the previous chapter I illustrated the ways in which art can form a critical inquiry of embodied post-conflict memory. This chapter complements the view by exploring the affective potential art can have for transformation and peacebuilding. In order to do this, I examine a Sámi art collective, *Suohpanterror* (Terror of Reindeer Lasso), and its activism. Bennett notes in relation to trauma-related art that art is often transactive rather than communicative. I interpret in this chapter the work of *Suohpanterror* to be a form of transactive art since it touches us, but "does not necessarily communicate the 'secret' of personal experience" (Bennett 2005, 7). The artists' work indeed derive from individual experiences as well as produce affect within and through the artwork, but the experiences and affects are accompanied with intellectual operations that stimulate critical thought which, in turn, is, in my view, a prerequisite for the transformation of violent relations.

The body that produces and responds to transactive art is visceral body. Visceral body is a feeling body that is also a biosocial body reflecting the interaction of biological and social forces. This body is inherently relational as bodily matter acts in its own right through its relationship to the social. Allison Haynes-Conroy and Deborah Martin (2010, 272) write that "the ways that bodies come to feel the world are reducible neither to cellular/chemical processes nor to discursive/intellectual processes; they are manifest in the catalytic relationship of the two". The visceral realm of the body is therefore multiple as the "body multiple" opens space for endless corporeal variations that include cellular/chemical and discursive/intellectual processes. Furthermore, the visceral body is constantly evolving in the sense that there is always space for new openings in this realm.

Visceral body and the touching capacity of art imply materialist ontology. As argued in Chapter 4 of this book in the spirit of Karen Barad's new materialism, touch is intra-action that is moving and affective. In concert with the new materialist ontology this chapter returns back to Jacques Rancière's understanding of politics introduced in the previous chapters. As noted, for Rancière politics is characterized by the de-classification of spatial and temporal orders. As also recognized earlier in this book, when speech is acquired by those whose right to speak is not recognized they can produce something new that disturbs the hegemonic representations of post-conflict identity constellations, namely they can evoke the questions "who are we?" and "where are we?" (Rancière 1999). In this view, politics is making visible the (colonial) boundaries and differences through which "our place and time" and "their place and time" are produced.

Slow Violence of Indigenous Governanance

Indegenous populations cannot be reduced under one homogenious category. The Sámi population has been decribed in the following manner:

> The Sámi are indigenous people of northern Fennoscandia. They are stereotypically associated with reindeer herding, but Sámi culture and lifestyle have varied greatly both geographically and chronologically.

> Historically, Sámi subsistence strategies were based foremost on fishing and hunting, sometimes in combination with agriculture, animal husbandry, and small-scale reindeer herding. Large-scale reindeer herding for meat production is a relatively recent phenomenon and is only performed by some Sámi. The Sámi of today are highly integrated into the societies of the nation states they live in, and, at least in the Nordic countries, they have relatively high levels of education, employment, and living standard, especially compared to many other indigenous populations in the world. (Spangen et al. 2015, 1)

The lifestyle which included hunting and seasonal migration changed when the authorities introduced settlement from the eighteenth century onwards. Settlement, on the other hand, provided means to impose a greater degree of control over the natural resources of the area and the Sámi population. Veli-Pekka Lehtola (2015, 29) notes that the economic utilization of northern resources proceeded hand in hand with national interests and the expansion of the state administration and institutions. Such institutions as the postal service, police, health service, and border guard were introduced. Schools were one of the major instruments of governing the population. The boarding schools were the Sámi children were enrolled due to long distances did not encourage the use of Sámi languages. On the contrary, it was forbidden to use the native language, even outside classrooms. Children spent long periods in boarding schools without visiting home and, thereby, the living experience of being a Sámi and belonging to the Sámi collective changed as shame of "not being a part" (Rancière 2004) became an integral part of the Sámi identity.

Furthermore, the Finnish nation practiced active politics of racialization. The 1920s marked the starting point for the Sámi racialization which aimed at "scientifically" verifying their lower racial quality. A part of the racialization was the myth of wild beast as Sámi people were exhibited in, for example, German zoos. Ristenrauna Magga narrates her experience of racialization:

> In autumn 1968 our family was invited to visit a school in Inari, and yes, we all packed ourselves to a car and travelled there. When we were there we were asked to undress. White coated men were measuring Sámi sculls and the length of our joints and limbs. Faeces and blood samples were collected. I remember that my joints were measured to be extraordinary long for a Sámi. (Ranta 2018)

When the erasure of native language is accompanied with racialization and these are embodied, as inevitably happens, they produce forceful collective affective landscapes. As indicated earlier, shame is a feature that has dominated the lived experience of being a Sámi. Shame has formed the "affective texture" of everyday life for many Sámi and this texture has been transferred from generation to another across time and space (see Wesslin 2017). However, shame does not permanently reside within bodies although for example the social and political formation of racialization creates shame as a long lasting component of collective life. As agued earlier, affect does not imply an individual feeling, but it is a "relation to others, a response to a situation and to the world", as Margaret Wetherell (2012, 24) argues. The corporeal affect of shame is a response to the colonial relations and violence embedded in them as affect is always intersecting, contextual, historical and interacting as well as changing.

As indicated above, settler colonialism functions through the practices of governance, e.g. forced settlement, schooling, taxation, and it establishes diverse techniques of the subjugation of bodies (Foucault 1978, 140). This type of power is "difficult to critically evaluate, protest against, or reform, precisely because it is a slow form of violence, often hidden out of sight" as Sara de Leeuw (2016, 16) writes in reference to the Canadian First Nations and settler colonialism. As noted earlier, colonial control mechanisms that regulate land, legal structures, territory as well as economic and political systems ultimately begin and end with bodies as they produce spatio-legal control and categorizations of indigenous bodies. According to de Leeuw, colonialism works through bodies and produces violence that is hidden and slow. Rather that creating spectacles and highly visual events of suffering, colonialism produces slow violence, i.e. "a tender, intimately-scaled, lived violence that is constantly being reproduced and reborn through individual bodies and relationships" (de Leeuw 2016, 17).

By unmasking settler colonial violence, the notion of slow violence reveals a temporal scale that produces gradual change. For Rob Nixon, who invented the notion, the Galtungian notion of structural violence is too static although it recognizes imperceptible violence. Structural violence is indeed a theory that entails rethinking different notions of causation and structurally embedded agency with respect to violent effects. For Nixon (2011, 11), on the other hand, as important as it is to recognize imperceptible violence it is to acknowledge "imperceptible change whereby violence is decoupled from its original causes by the workings of time" and "more complex descriptive categories of violence enacted slowly over time". Slow violence is characterized by the invisibility of the casualties; casualties that are not seen, counted or remembered. Nixon (2011, 13) writes about the casualties of slow violence that they "become light-weight, disposable casualties, with dire consequences for the ways wars are remembered, which in turn has dire consequences for the projected casualties from future wars". Furthermore, slow violence has, in my view, consequences how peacebuilding is enacted, if it is enacted at all given the invisibility of violence and vulnerabilities it creates.

Slow violence produces collective silence since it is often accompanied with a collective amnesia, and even with a full denial of the violence like in the Finnish case. There is nothing to be celebrated in the colonial relations and the abjected heritage as the relations remain "challenging" and "akward". As Elisabeth Edwards and Matt Mead (2013, 20) note, "narratives of the colonial past cannot be mobilized for 'positive didactic purposes'; nor can they be easily rehabilitated and reincorporated in the national imaginary", and yet they are constitutive of the colonizing nation. The colonial past and present resist being reduced to a coherent national narrative. Rather, they unfold "as an incomplete set of fluid relationships, as any number of unpredictable trajectories, over time and space" (Edwards and Mead 2013, 20).

Breaking the silence of colonial violence and moving towards peacebuilding is challenging. In many locations the reconciliation and indigenous rights is sought through truth commissions. The truth commission model is, however, limited in settler colonial relations as it is traditionally conceived as addressing the most serious human rights violations by the state or armed opponents after a conflict or an authoritatian regime. The experiences of Peru, Guatemala, and Chile show that there are multiple challenges: the comissions did not fully acknowledge the indegenous concerns, including the transmission on post-conflict memory; overt and sturctural violence were not sufficiently addressed; and the impact of different forms of violence were discussed partially. The Canadian experience, on the other hand, which dealt with the residential school system was an initiative of the native population itself directly engage with the state violence. In contrast to post-conflict or post-authoritarian truth and memory initiatives, this experience has forced the state structures face the the violation of indigenous rights and the denial of them. Although the commission did not fully remove the abjectification of the Canadian native people, it created an system of aknowledgement of the past violations. In Finland, the Finnish Government annouced in 2017 that it will establish a reconcialition process between the Finnish state and the Sámi population.

Colonial Appropriation of Bodies

Like war and violence that continue and end in bodies, colonial relations function in a similar manner. Bodies are used to represent and constitute the nation and its constitutive differences as nations translate biological bodies into social bodies in ways that reflect the particular collective project at that time. Kavita Daiya (2002, 235) writes that "women's bodies bear often the symbolic burden of community building—national or otherwise". She suggests that female bodies are appropriated for the purposes of locating the nation in the world, defining national boundaries and producing nation's enemies. However, as Daiya argues, an understanding of a nation must include attention to the ways in which both female and male bodies are appropriated and thereby suffer nationalistic, ethnic, gender and sexual violence (Daiya 2002). In this view, appropriation indicates a form of disciplinary power through which subjects are produced (Foucault 1973, 1979).

The Sámi art collective *Suohpanterror* explores these mechnisms of appropriation in its art. Among their works is a colourful poster that depicts the well-known Sun Maid raisin girl holding a plate full of grapes. The colours in the poster and of the text under it mimick the package of the Sun Maid raisins. The mimicking text reads: Mun Maid, Natural Sápmi Raisins. In the poster the Sun Maid girl's clothing is modified to resemble the Sámi female indegenous costume with its colourful neckerschief and a large silver brooch. In the photo of the poster there is a woman who is wearing a similar neckerschief and a large silver brooch than the modified Sun Maid girl in the poster. The woman is smiling and her her smile evokes a sense of irony, namely "look at our remseblances". The picture and the poster in the picture can be read to bring forth a seemingly innocent recemblance between a Sámi woman and the Sun Maid girl pictured originally in the famous Sun Maid raisins package. The colourful costumes, self-assured smiles and postures of these two women establish a performative connection that brings forth affective qualities of purity, originality, happiness and freshness. I, however, interpret the picture to make visible the commodification of racialized and gendered bodies since female Sámi bodies have been extensively used to market a variety of products and services.

There is a long tradition in Finland to promote, for example, tourism by using the images of (non-)Sámi—exclusively very light skinned and blond—women dressed in colourful costumes that imitate the traditional Sámi costumes. These images are presented as benign and non-political representations of the "exotic and authentic North" in order to attract tourists both from Southern Finlad and abroad to Lapland. A very recent version of expression of this promotion is a Facebook group titled *Lapland in words and pictures.* In 2015 the moderator of the group posted an old advertisement where two youg women are in front of a mountainous landscape, or a poster of the landscape, of the North wearing colourful Sámi dresses. The Finnish moderator of the group had named the picture in a patronizing tone to depict "two beautiful Sámi girls in their natural environment". The picture got a lot of thumps ups and the beaty and harmony of it were admired by many viewers. A Sámi activist who took part in the dicussion pointed out that the purpose of the picture is the commodification of the Sámi areas for touristic purposes and that the "Sámi girls" in the picture are Finnish models. Furthermore, the traditional looking costumes are not authentic traditional female Sámi costumes, but cheap copies of them. The moderator responded by stating that "why don't we all just enjoy the beauty of the image and leave dirty politics aside". He argued that the image should not be read in a political context as it is a genuine attempt to convey to the audience the serenity of the Northern landscape. The activist took pains to explain the long tradition of the misuses of the traditional Sámi costume as well as the commodification of the traditional costumes and their carriers. For her, this was an issue of identity, and ultimately it was about the right to keep ones cultural rights intact. The moderator of the group did apologise for commenting the picture, but maintained his right to post this kind of images in the future. He deleted the whole chain of discussion and stated new rules to the group. According to these rules, hate speech, insults, negative responses, taking part in Finnish-Sámi disputes, political opinions and racism are prohibited and will lead to a dismissal from the group.

The practices of colonial corporeal appropriation take multiple forms. In 2016, a Finnish bags and accessories manufacturer used in its marketing campaign an image of a woman wearing a fake traditional Sámi male hat. The company itself admitted the misuse and claimed that it just wanted to add the "exotic Sámi flavor" to its products to make them more attractive in the global fashion market. Sámi artists Marja Helander ja Outi Pieski made a strong statement against this type of colonial appropriation by writing in the largest Finnish daily paper:

> The Sámi costume, *gáhkti*, is one of the most important signs of the Sámi identity. It ties its carriers to the Sámi community. Symbolicly, *gáhkti* carries the wounds of the hundreds of years of repression and neglect. For the mainstream Finnish culture its just a colourful ornament and a means to market Finland. (Paltto 2016)

Similarly, colonial differences were established when a video titled *100 Days of Polar Night Magic* was released by the Finnish government's tourist agency, Visit Finland. In the video the imagery of colonial anthropology was evoked as the video presented dirty, wild, uncivilized and primitive Sámi dancing around a bonfire in a teepee. The uncritical reference by the agency to the tradition of colonial photography is highly problematic since the imagery referred to are expected to authorize and authenticate as well as offer unmediated windows on the world of the "primitivized other". As Edwards and Mead (2013, 24) note on this genre of imagery that it "itself becomes a site where the aesthetics of historiographical entanglements function to elide their colonial foundations and their temporal and cultural range" and that "style thus functions as another device through which to distance the colonial conditions [...]". The head of the marketing of Visit Finland, Merja Holmén, confimed that the video aimed at "introducing certain folklores and, yes indeed, we hope that these would interest our customers abroad. We hope to bring these folklores, including Sámi culture, forth in the future too" (Aikio et al. 2015). The video was disputed and the most explicit references to the "exotic and primitive Sámi" were finally cut off from it when the speaker of the Sámi Parliament Tiina Sanila-Aikio intervened and demanded that.

In sum, I argue that colonialism works through (gendered) bodies and practices targeted at them. The events and choreographies described above exemplify something what Albert Hurtado (1999) refers to as the "intimate frontiers" of empire, a social and cultural space where racial classifications are defined and defied, where relations between colonizer and colonized could powerfully confound or confirm the structures of governance and the categories of rule. The works of *Suohpanterror* and

Sami activism makes these frontiers visible by showing how the (mis)use of racialized and gendered Sámi bodies constitute the frontiers between the colonizer and the colonized. Yet, the voices of the colonized are seldom heard and the refusal of hearing constitutes a form of violence, namely epistemic violence.

Epistemic Violence

Feminist theorizing on linguistic competence and speech acts recognizes that power aids in ensuring that one's utterances receive the appropriate response, even despite the fundamental vulnerabilities all speakers have in linguistic exchanges. This view is applicable also to the situation of colonial relations that are always corporeal power relations as demonstrated earlier in this chapter. In Finland, the settler responses to the reasoning of the Sámi activists and the of work of *Suohpanterror* who question the appropriation of Sámi bodies, including the misuses of *gáhkti*, seek to patronize the "colonized child" whose reasoning is not valid and whose voice does not deserve to be heard. Jennifer Hornsby (1995) argues that in order speech acts to be successful there needs to be reciprocity. Successful linguistic exchange requires that the audience and speaker meet halfway, namely also the audience is willing to listen and hear. According to Hornsby's theory, the success of speaker's attempt to communicate ultimately depends on the audience. The dependency of the speaker indicates also her vulnerability because an audience may or may not meet the linguistic needs of a given speaker in a given event. I other words, the linguistic choreography remains one-sided if the needs are not met. Kristie Dotson (2011, 238) writes in reference to Hornsby's account that a "speaker cannot 'force' an audience to 'hear' her/him, where hearing refers to an audience fulfilling the demands for reciprocity in a successful linguistic exchange". In power relations such as settler colonial relations, the settler does not necessarily want to hear nor respond despite the fact that the colonized is able and wants to communicate. The colonizing audience is not response-able (Barad 2012, 206–207) as it constantly seeks to reject the subaltern's right to speak and, therewith, the audience refuses her knowledge claims.

One-sided linguistic exchange, or the lack of reciprocity, is a form of violence. Rob Nixon notes that different types of violence render different types of witnesses visible and leave others invisible. As demonstrated in Chapter 4, war veterans' witnessing becomes often acknowledged and recognized, but this does not happen with the indigenous witness voice. He (Nixon 2011, 16) writes in relation slow violence:

> To address violence discounted by dominant structures of apprehension is necessarily to engage the culturally variable issue of who counts as a witness. Contests over what counts as violence are intimately entangled with conflicts over who bears the social authority of witness, which entails much more than simply seeing or not seeing.

Being not recognized as witness is as such a form of epistemic violence. In the case of Finland's Sámi population, epistemic violence coincides with slow violence. For Gayatri Spivak, "epistemic violence" implies obstructing and undermining other than Western methods or approaches to knowledge. Ultimately, epistemic violence erases the space from which the subaltern could speak and where she could be the "knower". When the "knower" status is erased, Western forms of knowledge can be imposed on the marginalized groups. Through the violent dismissal of the other ways of knowing, and also speaking, colonial subject is constitutes solely as the inferior other and "different". Ultimately, according to Spivak, epistemic violence aims to alter the historical and social native consciousness as it seeks to delete all traces of the original and overwrite it with something considered more appropriate, i.e. with the linear and homogenous narrative and ways of knowing of, for example, the nation. Non-Western epistemology and ways of knowing the world are dismissed as inadequate and naïve as well as the reactions of the suppressed are deemed to be sentimental and exaggerated (Spivak 1994).

What characterizes epistemic violence is that the social group is understood according to stereotypes that strip it off from the ability to be identified as knower. The conversations where the Sámi activists seek to explain their traditions and claim them back, their identity as knower, and thereby also as witness of violence is denied. Their voices in conversations are rendered inferior. The Sámi activists' attempts to communicate their knowledge, for example about the usage of *gáhkti* and the importance of it for Sámi culture is seen to lack competence. A variety of "microinvalidation" strategies are initiated to dismiss the knowledge. Microinvalidatio is characterized by communications that exclude, negate, or nullify the psychological thoughts, feelings, or experiential reality of a person (Dotson 2011, 247). For example, the sentence "why don't we all just enjoy the beauty of the pictures" functions as an invalidation strategy that dismisses the lived and visceral experience of the speaker. Moreover, the strategies of microinvalidation function to dismiss entire collective identities as the Sámi case demonstrates.

Claiming the Colonized Body Back

In the settler colonial practices of appropriation, the colonized population is constituted outside, a frontier—a commodifiable and disposable frontier—in relation to the symbolic order of the nation-state. Michel Foucault (2000, 445) calls the practices that categorize and rank human bodies "intolerables" and argues that from them multiple points of resistance emerge. However, the frontier and the intolerables open up the possibility, if not the necessity, for the enactment of the political where politics composes itself through contingent, inventive and interruptive combinations. The work of *Suohpanterror* and Sámi activism is important as it resists the appropriation and symbolization, and as such becomes a remainder of the decades of slow and epistemic violence that escapes the linear and teleological identity narrative of the Finnish nation. Their activism and art present an existing interplay of political forces as they disrupt and destabilize and, thereby, restage the exclusion and colonial violence at the core of the Finnish nation. This can be seen to form in Jacques Rancière's (1999) sense an "enactment of the political" where politics emerges with respect to practices of governance and to different forms of violence.

There is space for resistance even in the most suppressive colonial relations. Foucault (1980, 142) writes about resistance that "there are no relations of power without resistances" and resistances are multiple, like the relations of power. In short, resistance is interlinked with power as they always appear together: power relations entail the possibility of resistance. Resistance to power can arise from within the very subject positions that power relations make possible. Resistance always implies in Foucauldian thinking specific and changing struggles in space and time. Aurelia Armstrong (2008, 24) summarizes Foucault's view on resistance:

> When Foucault tells us that in order to resist modern power structures we must not simply refuse identity, but actively perform that refusal through our attempts to produce new forms of subjectivity, he is once again evoking his conception of freedom as a practice which requires the active arrogation of the power of regulation by the subject for the purposes of self-formation or transformation.

For Foucault (1980, 216), the production of new forms of subjectivities requires self-creation. The aim of self-creation is not to discover what we are but to refuse what we are. Brent Pickett (1996, 462) writes with reference to Foucault that the "practice of the aesthetics of the self is nothing other than the resistance to the ways in which one is constituted as a subject of modern power".

Indigenous Sámi activist take part in the Foucauldian type of refusal of the colonial relations. One example of the refusal is Petra Laiti's performance in the Finnish National Gallery, Kiasma. In 2016 the Gallery exhibited a video where female figures dressed in latex underpants and in fake Sámi costumes dance and expose their butts. In Laiti's resistant performance three women entered to the gallery room where the video was displayed. They were dressed in real *gáhkti s* but wore them backwards. They sat silently in front of the video screen writing to paper slips their affective reactions to it. They wrote "I feel like crying", "I am tired" and "I feel like being abused" and filed them to museum's feedback box. An employee of the museum arrived and inquired whether the women had a permission to be in the room. Laiti reminisces the event and wonders: "Why is was just us who needed a permission to be there? There were also other people in the room, but they were not questioned. Just us who were dressed in Sámi costumes, only our presence there was questioned. That made us abjects". (Vasantola 2018)

Later Laiti was invited back to the museum to discuss the video and her performance with the museum's head and other representatives from the museum's administration. Despite Laiti's attempt to explain why the video was insulting, it was not removed from the National Gallery (Vasantola 2018). The director of the museum defended the video by claiming that the work discusses global commercial culture where different cultural traits are blended together. In a similar vein, the author of the video argued that her work had nothing to do with the Sámi (Laiti 2016). Despite the lack of recognition, acknowledgement and response-ability from the side of the colonizer, Laiti's performance made visible slow and epistemic violence that operates through material realities, through biopolitics of colonialism and, ultimately, through indigenous bodies.

Whilst Laiti claimed her appropriated body back, Niillas Holmberg claimed the entire Sámi nation back. In the afternoon of the 8th of October 2015 Holmberg, who was dressed in *gáhkti*, claimed up with a Greenpeace representative to the statue of the Finnish war marshal Carl

Gustav Mannerheim at the centre of Helsinki. They attached "Save the Artic" poster to one side of the statue and "Self-determination" to the other. Holmberg had a Sámi flag and he was waiving it in the cool afternoon sun. At one point of the performance Holmberg was wearing a balaclava in Sámi colours. Earlier in the same morning Finnish authorities had discussed the future of the Finnish natural resources, and particularly forestry. In Finland, the state is one of the biggest single forest and land owners, and that has caused many land and resource related conflicts between the Sámi and the Finnish state. A new law concerning the land and water resources was written at that time and the Sámi argued that their views had not been taken into account (Häirikkö blog 2015). Holmberg commented the performance and the preparation of the new law by stating that the "Finnish state is exploiting the natural and cultural resources of the Sámi population" (Kärki 2015).

Holmberg's choice of the location of the performance holds a great symbolic value. In Finland, war marshal Carl Gustaf Mannerheim has been encoded to represent the ultimate heroism and honor in the Finnish state. Alternative ways of encoding him has been largely shunned away in the Finnish society. He was born in 1867 and spent the first period of his military career in the Russian army. He served as supreme commander of the Finnish armed forces in three wars, the Civil War of 1918, the Winter War of 1939–1940 and the Continuation War of 1941–1944. He was also President of Finland in 1944–1946. His "experience, character and physical presence are set up as metonymic of the national", as Elleke Boehmer (2005, 69) writes about heroic male figures and the ways they are constituted to represent the nation. In this linear narrative his past becomes the true past of the Finnish nation. The narrative does not allow alternative histories to emerge: it is the grand history that dismisses the "other", e.g. indigenous, history.

These Sámi performances force to re-think the nation's identity politics since they make visible "a liminal signifying space" that is marked by the heterogeneous histories of contending people and the tense locations of political differences (cf. Bhabha 1994, 148). In the postcolonial and subaltern studies, there is an abundance of scholarship that shows the limits of hegemonic history writing and the idea of the unitary nation (Bhabha 1994; Chakrabarty 2000; Guha 1999; Spivak 1999). As argued in the previous chapters, national history works to secure for a contingent nation the imaginary unity of the national subject evolving through time. According to Prasenjit Duara (1995, 29), "the subject of History is a metaphysical unity devised to address the aporias in the experience of linear time" . However, the favouring of one past which is claimed to be true is not without its fissures as there is a constant pull towards heterogeneous histories that threaten to bring difference and contestation to the core of the nation. By claiming the female Sámi body back and by occupying the statue of Mannerheim and demanding self-determination, the performances made difference and contestation visible.

Violent colonial past is not simply Laiti's and Holmberg's personal experience, but it is a part of a transversial historical process that they have shared with of other colonized people. The performances demonstrate that many knowledge claims, and thereby truths, exist in parallel and some of them have been wiped out due to their disturbing qualities. Politics in theses cases is created through a moment of interruption, a moment where the Sámi presence disturbs the national order of things. The fusing of the national "true" past with the Sámi present causes the resurgence of difference and contestation that destabilizes the already established non-colonizing national identity. A discontinuity emerges between the nation's present status and its myth of being a singular non-colonial subject.

The performances can be understood hence as material and a symbolic time-place events that shatter the hegemonic distribution of the Finnish settler colonial history and its "truth" and knowledge claims. They enact the wrong and expose the ruptures that cut through the national body politic. They introduce political dissensus that is not a discussion between speaking people who would confront their interests and values or who disagree with the presence of a common object, but rather it concerns the very capacity to the interlocutors to present the object as indicated earlier in reference to Ranciére's (1999) theorizing. It brings forth the political where the political involves a conflict about who speaks and who does not speak, about what has to be heard as the voice of pain and what has to be heard as an argument on justice.

Summary

I have demonstrated in this chapter that nations do not always want to recognize the settler colonial relations at the core of them, and thereby long-term peacbuilding is impossible. Slow and epistemic violence that function though the indigenous bodies render the bodies mute from the perspective of the state. What is needed, in my view, is the enactment of the political, namely a capacity to evoke the question who belongs to the national and political community, where are the frontiers of the community and who is cast out and how the casting out is done.

Political struggle in the settler colonial context is not a conflict between well-defined interest groups; it is an opposition of logics that count the parties and parts of the community in different ways. Rancière (2001, thesis 7) writes about the political struggle:

> The essence of politics, then, is to disturb this arrangement by supplementing it with a part of the no-part identified with the community as a whole. Political litigiousness/struggle is that which brings politics into being by separating it from the police that is, in turn, always attempting its disappearance either by crudely denying it, or by subsuming that logic to its own. Politics is first and foremost an intervention upon the visible and the sayable.

It is in this sense the performances of the Sámi activists introduce political, settler colonial, struggle to the core of the nation. They do not rupture the power positions as such, but rather challenge the classifications of those worthy of inclusion and those excluded as well as bring forth a space for the appearance of a colonized subject. They engage in a radical political practice that displaces the limits and frontiers of social exclusion. Political struggle in this sense is not a rational debate between multiple interests, as noted earlier, but the struggle for one's voice to be recognized as the voice of a legitimate partner.

References

Aikio, K., Länsman, K., Alajärvi, M., & Aikio, A. (2015). Likaiset lappalaiset, sekö myy maailmalla?. YLE 17. 9. 2015. Accessed 30 May 2018. https://yle.fi/uutiset/3-8312073.

Armstrong, A. (2008). Beyond Resistance: A Response to Žižek's Critique of Foucault's Subject of Freedom. *Parrhesia, 5*, 19–31.

Barad, K. (2012). On Touching—The Inhuman That Therefore I Am. *Differences, 23*(3), 206–223.

Bennett, J. (2005). *Empathic Vision*. Standford: Standford University Press.

Bhabha, H. (1994). DissemiNation: Time, Narrative, and the Margins of the Modern Nation. In H. Bhabha (Ed.), *Nation and Narration* (pp. 139–170). New York: Routledge.

Boehmer, E. (2005). *Stories of Women: Gender and Narrative in the Postcolonial Nation*. Manchester: Manchester University Press.
Chakrabarty, D. (2000). *Provincializing Europe: Postcolonial Thought and Historical Difference*. Princeton: Princeton University Press.
Coulthard, G. (2007). Subjects of Empire: Indigenous Peoples and the "Politics of Recognition" in Canada. *Contemporary Political Theory, 6*, 437–460.
Daiya, K. (2002). "Honourbale Resolutions": Gendered Violence, Ethnicity, and the Nation. *Alternatives, 27*(2), 235.
De Leeuw, S. (2016). Tender Grounds: Intimate Visceral Violence and British Columbia's Colonial Geographies. *Political Geography 52*(2), 14–23.
Dotson, K. (2011). Tracking Epistemic Violence. *Tracking Practices of Silencing. Hypatia, 26*(2), 236–257.
Duara, P. (1995). *Rescuing History from the Nation: Questioning Narratives of Modern China*. Chicago: University of Chicago Press.
Edwards, E., & Mead, M. (2013). Absent Histories and Absent Images: Photographs, Museums and the Colonial Past. *Museum & Society, 11*(1), 19–38.
Facebook Group. *Lapland in Words and Pictures*.
Foucault, M. (1973). *The Birth of the Clinic: An Archaeology of Medical Perception*. London: Routledge.
Foucault, M. (1978). *The History of Sexuality Volume 1: An Introduction*. New York: Vintage Books.
Foucault, M. (1979). *Discipline and Punish: The Birth of Prison*. New York: Vintage Books.
Foucault, M. (1980). *The History of Sexuality Volume 1: The Will to Knowledge*. London: Penguin.
Foucault, M. (2000). *Power: Essential Works of Foucault, 1954–1984*. Vol. 3. Ed. Paul Rabinow. New York: The New Press.
Guha, R. (1999). The Small Voices of History. In S. Amin & D. Chakrabarty (Eds.), *Subaltern Studies IX, Writings on South Asian History and Society* (pp. 1–12). New Delhi: Oxford University Press.
Häirikkö Blog 8. 10. 2015. Suohpanterror iski Helsingissä. *Voima Magazine*. Accessed 30 May 2018. http://voima.fi/blogikirjoitus/2015/suohpanterror-iski-helsingissa%E2%80%A8/.
Hayes-Conroy, A., & Martin, D. (2010). Mobilizing Bodies: Visceral Identification in the Slow Food Movement. *Transactions, 35*, 269–281.
Helsingin Sanomat, 6.8.2017, section B1–B5.
Hornsby, J. (1995). Disempowered Speech. *Philosophical Topics, 23*(2), 127–147.
Hurtado, A. (1999). *Intimate Frontiers: Sex, Gender, and Culture in Old California*. Albuquerque: University of New Mexico Press.
Kärki, J. (2015). Aktivistit iskivät: Mannerheimin harteille Saamen lippu. Accessed 30 May 2018. https://www.mtv.fi/uutiset/kotimaa/artikkeli/aktivistit-iskivat-mannerheimin-harteille-saamen-lippu/5351320#gs.I1IFYcU.

Lederach, J. P. (1997). *Building Peace: Sustainable Reconciliation in Divided Societies*. Washington, DC: U.S. Institute of Peace Press.
Lehtola, V. P. (2015). Sámi Histories, Colonialism, and Finland. *Artic Anthropology, 52*(2), 22–36.
Morgensen, S. L. (2011). *Spaces Between Us: Queer Settler Colonialism and Indigenous Decolonization*. Minneapolis: Minnesota Press.
Nixon, R. (2011). *Slow Violence and the Environmentalism of the Poor*. Cambridge, MA: Harvard University Press.
Paltto, A. S. (2016). Kiasmaa arvostellaan feikkisaamenpukuvideon ostamisesta. Accessed 30 May 2018. https://yle.fi/uutiset/3-8869765.
Petra Laiti, Blog 2016. Meanwhile in Kiasma. Accessed 30 May 2018. https://petralaiti.com/2016/09/10/meanwhile-in-kiasma/.
Pickett, B. (1996). Foucault and the Politics of Resistance. *Polity, 28*(4), 445–466.
Rancière, J. (1999). *Disagreement*. Minneapolis: University of Minnesota Press.
Rancière, J. (2001). Ten Theses on Politics. *Theory and Event, 5*(3). Accessed 1 Oct 2014. http://www.egs.edu/faculty/jacques-ranciere/articles/ten-thesis-on-politics/.
Rancière, J. (2004). *The Politics of Aesthetics: The Distribution of the Sensible* (Gabriel Rockhill, Trans. and Introd.). London and New York: Continuum.
Ranta, K. (2018). Ristenrauna Maggan perhettä nöyryytettiin kallonmittauksilla vielä 1960-luvulla. YLE 19. 2. 2018. Accessed 30 May 2018. https://kukkaranta.com/2018/02/19/ristenrauna-maggan-perhetta-noyryytetti-in-kallonmittauksilla-viela-1960-luvulla-saamelaiset-toivovat-totuuskomission-tuovan-julki-menneisyyden-vaaryydet/.
Spangen, M., Salmi, A. K., & Äikäs, T. (2015). Sámi Archaeology and Postcolonial Theory—An Introduction. *Artic Anthropology, 52*(2), 1–5.
Spivak, G. (1994). Can the Subaltern Speak? In C. Nelson & L. Grossberg (Eds.), *Marxism and the Interpretation of Culture* (pp. 275–313). Chicago: University of Illinois.
Spivak, G. (1999). *A Critique of Postcolonial Reason: Toward a History of the Vanishing Present*. Cambridge, MA: Harvard University Press.
Vasantola, S. (2018). Twitterin päivystävä saamelainen Petra Laiti: Saamelaisten sorto pahenee yhä. Helsingin Sanomat 6.1.2018. Accessed 30 May 2018. https://www.hs.fi/sunnuntai/art-2000005514679.html.
Veracini, L. (2013). Introducing Settler Colonial Studies. *Settler Colonial Studies, 1*(1), 1–12.
Wesslin, S. (2017). Kielensä kadottaneet lapset – näin saamen kieli pestiin pois. YLE 7.5.2017. Accessed 30 May 2018. https://yle.fi/uutiset/3-9579558.
Wetherell, M. (2012). *Affect and Emotion: A New Social Science Understanding*. London: Sage.

CHAPTER 8

Conclusions: Corporeal and Response-able Peacebuilding

Abstract The chapter summarizes the theoretical argument presented in the book, namely why the turn to the sentient body is valuable for the sake of vitalizing the study of peacebuilding and peace. The events and narratives of ordinary citizens, war veterans, abjected subjects and indigenous activists on which the empirical material of the book has built its arguments speak persuasively of the richness of the body as a site and source of the political of post-conflict, agency, conformation and resistance, and thereby of peacebuilding. Bodies pose a challenge to the study of peacebuilding as they force to reconsider the content of post-conflict political community and the instances of visibility, acknowledgement, recognition and belonging. They invite to create response-ability, that is a prerequisite for sustainable peacebuilding.

Keywords Corporeal event · Response-able peacebuilding

I have urged in this book the appreciation and look for the value of bringing the body, the everydayness and ordinary people to the study of peacebuilding. Since my aim has been to bring political theory, feminist theorizing and post-colonial theory to Peace and Conflict Studies the concluding chapter is not to summarize the case studies, but to discuss again peacebuilding from a theoretical perspective based on the above-mentioned corpus of literature that brings together the findings of the case studies. The move, or turn to the body, is valuable for the

T. Väyrynen, *Corporeal Peacebuilding*, Rethinking Peace and Conflict Studies, https://doi.org/10.1007/978-3-319-97259-6_8

sake of vitalizing the study of peacebuilding and peace, instead of conceiving them merely as an abstraction. I have outlined in the book how bodies pose a challenge to peacebuilding that is both praxical and philosophical: they force us to reconsider the content of post-conflict political community, recognition and belonging. These are all ultimately questions about post-conflict social and political formations, identifications and agency. The question in this book has hence been one of how new forms of post-conflict being-in-the world that depart from the old order might be produced through peacebuilding. In the domain of epistemology and ontology, the question is then no longer that of the nature of peacebuilding "reality", but rather of the distribution of visibility, and thereby recognition, acknowledgement and belonging.

The case studies of the book can be read as examples of corporeal events that rupture normality and, at the same time, open a space to re-think peacebuilding arrangements and thereby future possibilities for peace. Thinking through event-ness and the choreographies of events directs Peace and Conflict Studies towards the fleshy and carnal existence from which the mundane practices of peacebuilding emerge and that are constitutive of long-term and sustainable peace.

By attesting to the body through different ontological positions enables Peace and Conflict Studies to explore how an experience of conflict and peacebuilding are experiences of pervasive presence that gives shape to one's sense of direction, place, space and time of conflict and peacebulding. Within such a framework, the experiences of violence and peacebuilding are inextricably linked to the rhythm of the body that may or may not coincide with the framework of the institutions of war and peacebuilding create. This opens up the question of relation between bodies, the experience of being in the presence of others and the rhythm and choreography of negotiating between the senses of the world as bodies resonate with others, in relation to others. Furthermore, it invites to conduct a corporeal analysis, that brings together affect, emotions and the somatic and to provide an understanding of the body as both the subject and object of discourses, practices and policies of peacebuilding.

The study of conflict and peacebuilding as socio-political institutions encourages the examination of anybody who has experiences and agencies that are related to violence, conflict and war as these are corporeally carried over to peacebuilding. In other words, conflict and peacebuilding are not only about soldiers, statesmen, international institutions, victims and heroes. It is equally important to shed light on phenomena that are seldom examined in Peace and Conflict Studies, which include trans-generational memory, resistances, silences, post-conflict governance of subaltern bodies, the affective dimensions of war and conflict and senses. All these are corporeal phenomena. By focusing on the corporeal, temporal and spatial transitions, this type of examination of war and peacebuilding can avoid reductionist interpretations that shy away from seeing the practices through which differences, frontiers are silences are produced in peacebuilding. The emphasis in this book has been on the mundane and comprehensive experiences of existence at a time of conflict and peacebuilding and, thereby, the chapters have shed also light on the multiple forms of bodily agency, some of which are quite controversial and disturbing for the post-conflict political order.

The approach in this book opens up the relationship between peacebuilding, space, time and the body in novel ways, which, in turn, enable to tweak the notion of peacebuilding from being a statist endeavor towards understanding the political that always already lies with the body. The events and narratives of ordinary citizens, war veterans, abjected subjects and indigenous activists on which the empirical material of the book has built its arguments speak persuasively of the richness of the body as a site and source of the political of post-conflict, agency, conformation and resistance and, therewith, of peacebuilding. The reading of the narratives, visual material and performance art presented in the book underscores that even though the body may constitute an important and intriguing topic within Peace and Conflict Studies, it always defies all attempts of comprehensive categorization and exhaustive reading. Focusing on the body reveals the relational and political—and messy—nature of peacebuilding. As indicted in Chapter 1, peacebuilding and peace are dynamic corporeal events that do not take place just "there and then" but, but also "here and now" when variously located bodies bring the touch of war home, make epistemic and slow violence visible, speak back to power and remind the community of the post-generational transition of violence and conflict.

Through attending to mundane, relational and sensing bodies Peace and Conflict Studies can appreciate that bodies are embedded in history and as such they are sites of collective memory. War, political oppression, torture, atrocity, catastrophe and societal breakdown are remembered and incorporated in the body to an extent that they become worn on the body. It follows that experiences, narratives and memories of violence and conflict reach always beyond personal reflection and individual memory. To study conflict and peacebuilding through such a spatio-temporal spectre makes it meaningful to consider how the nexus of conflict and peacebuilding takes shape on bodies and also the ways in which bodies respond to, make sense of, and resonate with conflict and peacebuilding. Through empirical case studies that attend, for example, to witnessing, wounded, remembering, silenced and resistant bodies it is possible to examine how war and violence are transferred across time and space to other bodies, from one generation to another and across spatial surfaces to peacebuilding as well as emphasize that peacebuilding is not a separated event that can bring forth a political and societal closure, but an ongoing political entanglement that has to be constantly re-configured.

My take on the nexus between the body and peacebuilding has underscored that the singular sovereign subject is an impossible concept, as we only exist in relation, with one another. The relationality of human being holds relevance also to peacebuilding as the "truth" of experience lies between bodies, and as such it can never be fully (dis)closed and concluded by one body alone. The book has also demonstrated that the body and sense tell much more about rhythm and corporeal arrhythmia, or ambiguity, messiness and openness of peacebuilding than the stock theories and conventional (national) histories of conflict and peacebuilding, which are often devoid of everybody. The response to the messiness and openness of peacebuilding is not consensus, but on the contrary, dissensus. In the Rancèrian reading, corporeal peacebuilding necessitates dissensus that is not a discussion between speaking people, but rather it concerns the very capacity to the interlocutors to be a part of the discussion. In the case of peacebuilding the stakes are high since it involves always a struggle who has the right to present post-conflict community, who has right to be visible and who belongs to the community. As corporeal experiences bring constantly forth the other pasts and presences, relationalities and vulnerabilities, post-conflict peacebuilding is ultimately about the emergence of dissensus that concern these.

The assumption of this book has been that ontology is not singular and given and, thereby, the question arises about which reality to "do" as there are always specific relevant interventions to be made. In this view, ontologies do not precede or escape politics, but have politics of their own. The question is about a politics of what is the reality that takes shape and that various people come to live with. This approach offers a novel perspective to peacebuilding as it deals with the corporeal reality of peacebuilding. Rather than asking whether or not some ontologies know the body correctly, the book has shifted to the question what are the topics, the concerns and the questions that these ontologies bring forth in relation to the body, conflict resolution and peace, i.e. corporeal peacebuilding. Furthermore, as there is more reality than one, its different versions are variously entangled with one another that can be examined from multiple angles, each of which offer a slightly different, yet complementary, understanding of peacebuilding.

The turn to the body and corporeal agency brings unavoidably forth ethics. Karen Barad's new materialism demonstrates that it is the very nature of one's exposure to the others, vulnerability, that brings forth responsibility and, thereby offers a new, *response-able*, take on peacebuilding. The responsibility that follows from the exposure is not voluntary, but rather an incarnate relation that is embedded in the process of intra-action and world's ongoing intra-active becoming. Barad (2012, 217) writes about our exposure to the outside that "this sense of exposure to the other is crucial and so is the binding obligation that is our vulnerability, our openness [...]". Being in a relationship with another is a matter of being outside, of "being lived through by forces that constitute the human subject" (Barad 2012, 218). When we are touched, for example, by bodies that have experienced violence and conflict, we live through the (violent) forces that constitute them. These bodies are penetrated by the alterity of violence, and this calls us inevitably to a new sensibility, to "com-passion – suffering together with, participating with, feeling with, being moved by" (Barad 2012, 216) as material bodies touch and resonate with each other. Resonating with another body (matter) is coming in contact with something in a way that enables *response-ability*. In short, the body brings forth also the realm of ethico-political as it invites response. Response is an unavoidable part of peacebuilding since through it acknowledgement, recognition and visibility are allocated.

I have demonstrated in the book that materialist ontology has implications also for thinking about peacebuilding agency as it enables "interrogation of mythologies of liberal personhood and sovereign agency by foregrounding the human body's autonomic 'prior-ness' to the social world" (Tompkins 2016). Bodies come to matter through world's intra-activity: bodies arise intra-actionally where their momentary boundaries are produced. Matter, including bodies, is not sheer physicality nor is it immutable or passive. It does not require the mark of an external force like culture or history to complete it as it is always already an ongoing historicity. This view enables new avenues for corporeal peacebuilding by dissolving the atomic nature of the self. By dissolving the boundaries of the self and other, Peace and Conflict Studies can think through bodily affect as collective and social, which is, at the same time, ethico-political move towards collectivization and new forms of peacebuilding agency.

References

Barad, K. (2012). On Touching—The Inhuman That Therefore I Am. *Differences, 23*(3), 206–223.

Tompkins, K. (2016). New Materialisms. *Lateral, 5*(1). Accessed 31 May 2018. http://csalateral.org/issue/5-1/forum-alt-humanities-new-materialist-philosophy-tompkins/.

Index

T. Väyrynen, *Corporeal Peacebuilding*, Rethinking Peace and Conflict Studies, https://doi.org/10.1007/978-3-319-97259-6

The manufacturer's authorised representative in the EU is Springer Nature Customer Service Centre GmbH, Europaplatz 3, 69115 Heidelberg, Germany. If you have any concerns regarding our products, please contact ProductSafety@springernature.com

Printed and bound by CPI Group (UK) Ltd, Croydon, CR0 4YY
22/09/2026
02221067-0014